RECENT EARTH HISTORY

RECENT EARTH HISTORY

Claudio Vita-Finzi

University College London

MACMILLAN

First published 1973 *by*
THE MACMILLAN PRESS LTD
London and Basingstoke
Associated companies in New York Dublin
Melbourne Johannesburg and Madras

SBN 333 15027 9 (hard cover)
      333 15136 4 (paper cover)

Printed in Great Britain by
A WHEATON & CO
Exeter

To P

Contents

Acknowledgments

I am greatly indebted to the following, as well as to the authors concerned, for permission to use illustrative and other material : W. H. Freeman and Co. (figure 2); The Geological Society of America (figures 4, 9B, 11 and 12D); The University of Colorado (figure 5); Almqvist and Wiksell Förlag AB (figure 6); The M.I.T. Press (figure 8); *American Journal of Science* (figure 9A); *Botaniska Notiser* (figure 10); The American Association for the Advancement of Science (figure 11A and 13 : Copyright 1968 and 1964 by the American Association for the Advancement of Science); UNESCO (figure 11A); Longman Group Ltd (figure 12A); *Quaternaria* (figure 12B); Macmillan (Journals) Ltd (figure 12C); The University of Chicago Press (figure 14C : Copyright 1955 by the University of Chicago); Jonathan Cape Ltd (figure 14B); *Marine Technology Journal* (figure 15); Gebrüder Borntraeger (figure 16B); The University of Texas Press (figure 17A); Faber & Faber Ltd (figure 18); The American Association of Petroleum Geologists (extracts from the Code of Stratigraphic Nomenclature in chapter 2 and 7).

Preface

This book was conceived as an elephantine compendium and has emerged as a polemical mouse. My original intention was to write an account of the physical changes undergone by the earth during Holocene (or Postglacial) times, in the form of a stratigraphical narrative embellished by a few key radiocarbon and historical dates. But, as the work proceeded, I became convinced that dating, rather than serve as an adjunct to stratigraphy, deserved to replace it as the basis of chronology. Some will find this view self-evident, but as it did not appear to be widely held by geologists and prehistorians I decided to argue its merits using the last 20 000 years as the main source of illustrations.

It may seem fraudulent to base the general case on a part of the record which is exceptionally well endowed with quantitative dates. But my aim is simply to demonstrate the benefits of a particular approach : dentifrice advertisements do not usually show toothless gums. Moreover, if the proposals are at present difficult to implement throughout the earth's history it does not follow that the underlying principles are fallacious. In any case many of the recent dates already on hand have been little exploited, whereas much of what is preached here is already common practice among students of the Precambrian, if at times only because a lack of fossils or of stratification debars them from more conventional methods of subdividing their material.

As the book presents a very definite point of view it may do less than justice to studies which do not conform. The inclusion of detailed references and notes to each chapter will partly atone for this; nevertheless the intention has been to make the text self-sufficient. The first three chapters deal with the problems of dating and correlation in general. Chapters 4 to 8 consider how quantitative dating can help in the study of selected sequences, whether through the 'metering' of local changes or the construction of regional time-planes and time-lines. The final chapter attempts to show that such procedures lead on to productive generalisation. I should add that, although physical geology dominates the case-studies (with prehistory in second place), I have tried to make the book palatable to readers primarily concerned with other aspects of terrestrial chronology.

What originality there is in the pages that follow stems from studies I was enabled to carry out by the generous and tolerant support of the Nuffield Foundation. I am also indebted to John Adams, Eric Higgs, David Krinsley, John Pfeiffer and Michael Wood for their comments on parts of the text, and to Alick Newman and Rick Bryant for drawing the figures.

1 Introduction

It is not possible to dig a hole in a different place
by digging the same hole deeper.
Edward de Bono, *The Use of Lateral Thinking*

To judge from the prominence given to 'absolute dates' by earth
scientists, archaeologists and workers in allied fields, age determination has come to play an essential part in the chronicling of the
history of the earth and its inhabitants. The flourishing state of
research into existing and novel dating techniques guarantees
further progress in this direction; and progress it undoubtedly
represents to anyone curious of the past. Yet many of the principles
on which geology and prehistory rest were formulated at a time
when techniques for measuring the age of rocks and fossils were
almost wholly lacking. Do they need reassessment now that quantitative dating is practicable beyond the historical period?
Geochronology has been defined as the science of dating events
in earth history. In his classic *Dating the Past*, F. E. Zeuner stressed
the close relationship between geochronology and stratigraphy, and
pointed out that all the methods he would consider relied upon
strata of some kind or other.[1] The methods included radiocarbon
and potassium–argon dating as well as those that involve the
counting of tree-rings, varves and other 'stratified' items.[2] More
recently, R. F. Flint has defined the role of geochronology as the
'calibration of a time-stratigraphic sequence', claiming that stratigraphy and chronology taken together provide the framework upon
which earth history is constructed.[3] The alliance is not peculiar to
the Quaternary period (figure 1)—the main concern of both these
authors—it being generally accepted that any 'absolute geologic
time scale must be built up by interpolating between suitable events

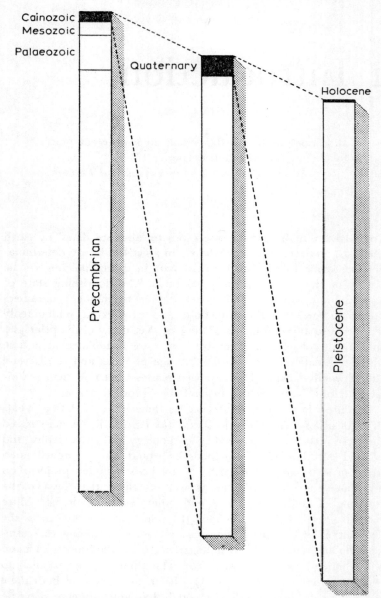

Fig. 1 The Quaternary period and its subdivisions in relation to the
geological time scale.

that can be not only dated in absolute terms but also accurately placed in the relative time of stratigraphy'.[4]

The benefits that have accrued from calibrating existing stratigraphical units are considerable and well known. But it is still legitimate to ask whether something might be gained from occasionally disengaging dates from the existing framework.[5] (The fact that many of the available age measurements cannot be fitted unequivocably into the stratigraphical system would in itself seem sufficient motive for making the experiment.) Paradoxically, the greatest support for such a move is given by those stratigraphers who equate geochronology with biochronology, that is to say with palaeontological dating, their chief objection to the radiometric time scale being that it is totally independent of the geological time units derived from geological history.

But this is to confuse the issue. Let us simply accept that, as de Cserna has recently found it necessary to point out, whereas rocks are measured in metres and wine in litres, time is measured in days or years.[6]

2 Stratigraphical earth history

I deny, in a high number of instances, the existence
of succession. I deny, in a high number of instances,
contemporaneity as well.

Jorge Luis Borges, *A New Refutation of Time*

We are frequently told that the history of the earth is written in
the strata of its crust. The metaphor gains in felicity the more it
is developed : the geological eras represent successive volumes; the
periods, chapters; and rock succession at any one place can be
likened to 'an Egyptian chronicle containing only a few years at
a time, taken at random from different dynasties'.[1]

Stratigraphical concepts undoubtedly underlie the writing of
earth history. Before considering their present status we must briefly
review their evolution.

Stratigraphy and geology

Historical geology has antecedents which can be traced into early
Antiquity. Origen, better known in another context,[2] quotes
Xenophanes (*fl.* 540 B.C.) to the effect that the earth is prone to
solution by the sea, witness the presence of marine shells far inland.
Xanthos of Sardis (*ca.* 500 B.C.) used similar evidence to show that
the sea once covered districts now dry. Many other classical authors
could be cited who showed an awareness of geological change, and
they were followed by countless observant and imaginative writers.
But not until the late eighteenth century was it 'discovered that

the rocky strata of the earth contained a succession of records which, if deciphered, would present a history of the earth from its beginnings to the present day'.[3]

Steno, who published a celebrated study of Tuscany in 1669, had shown that rocks could take the form of superimposed layers, that the order in which the layers occurred was repeated in different exposures, and that in undisturbed sedimentary deposits the lower layers were older than those above. These points were restated and amplified by later workers, among them James Hutton; others noted that fossils sometimes changed in character from one stratum to the next. The science of stratigraphy arose from the fusion of the physical with the organic record by Cuvier, Brongniart and, above all, William Smith (1769–1839).

At first, fossils were employed as an adjunct to rock type in tracing strata from place to place. It was then found that distinctive fossils or fossil assemblages could be used in their own right to link beds differing in composition and to establish the relative age of beds which were not juxtaposed. The principles of stratigraphy have since come to permeate historical geology as a whole. Even where stratified, fossiliferous rocks are not the primary concern, the aim usually remains that of establishing local sequences which can then be combined by correlation to yield a regional account. The sequences may thus include episodes of crustal deformation, changes in the orientation of the earth's magnetic field, or any other item that ranks as a geological event.

Stratigraphical practice

Several codes of stratigraphical practice are in existence, and attempts are being made to devise a scheme that will prove internationally acceptable.[4] Three kinds of units are widely regarded as essential to stratigraphy; the definitions given here are drawn from the code prepared by the American Commission on Stratigraphic Nomenclature in 1961.

A rock-stratigraphic [lithostratigraphic] unit is a subdivision of the rocks in the earth's crust distinguished and delimited on the basis of lithologic characteristics. . . . Concepts of time-spans, however measured, properly play no part in differentiating or determining [its] boundaries. . . .

A biostratigraphic unit is a body of rock strata characterized by its content of fossils contemporaneous with the deposition of the strata.

A time-stratigraphic [chronostratigraphic] unit is a subdivision of rocks considered solely as the record of a specific interval of geologic time.[5]

The crucial distinction would seem to lie between the observable criteria of lithology and palaeontology, and the 'inferential' basis of chronostratigraphy.

Correlation has been defined as the demonstration of equivalency of stratigraphical units.[6] Consequently, three kinds of correlation can be recognised, namely rock-stratigraphical, biostratigraphical and time-stratigraphical, of which the first two may be coupled under the heading of rock-correlation in order to sharpen the contrast with time-correlation. Some workers use correlation on its own to imply rock-correlation, time-correlation or a bit of both; in the definition given above, 'equivalency' is no less elastic. Others have proposed that correlation be used solely to denote time-equivalence, but this has been criticised as a rather severe restriction on a useful general term.[7] 'All English-speaking stratigraphers know very well what they mean by correlation', an eminent stratigrapher remarks,[8] going on to cite the second definition given in *The Oxford English Dictionary* (*13 vols., 1933*) :

verb 2. To place or bring into correlation; to establish or indicate the proper relation between (*spec.* geological formations, etc.).

The illustration given in the OED very properly comes from a geological work, Murchison's *Siluria* (1849).

For the newcomer to the subject, rock-stratigraphical correlation presents the least mystery : once a unit has been defined in terms of its physical characteristics, its counterpart will presumably be sought in other sections or boreholes. The volcanic ashes emitted by a shortlived eruption, for example, may be so distinctive as to be traceable over thousands of square kilometres. But, as most rock-stratigraphical units change in character laterally, their recognition will come to depend on their position relative to more persistent items, on their fossil content, and the like. Indeed, there are many stratigraphers, notably in Great Britain, who follow

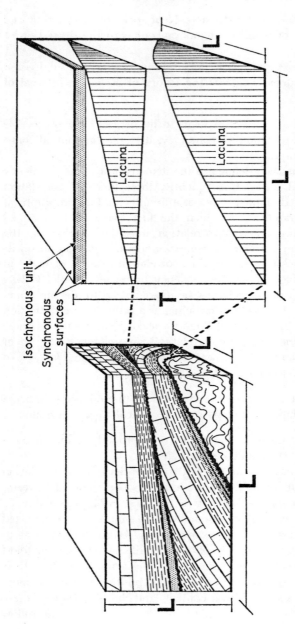

Fig. 2 Strata and time. The block diagram on the left depicts a stratigraphical sequence which includes two unconformities (wavy lines) due to nondeposition, erosion or both. All three dimensions are linear (L). The block on the right is an area–time diagram in which the vertical dimension represents geological time (T). The white zones denote the period spanned by the deposition of the rocks preserved at different localities; the zones shaded vertically correspond to the unconformities of the first figure. An isochronous stratigraphical unit is shown for comparison. Adapted from *Stratigraphy and Sedimentation*, Second Edition, by W. C. Krubein and L. L. Sloss. W. H. Freeman and Company. Copyright © 1963.

William Smith in defining formations by a combination of litho-logical and palaeontological features; in the American code, how-ever, the formation is the basic lithostratigraphical unit.

Although recourse may be had to ancillary sources, if not to intuition, in correlating biostratigraphical units, these generally live up to their name in being based on palaeontological criteria. On the assumption that evolutionary change is irreversible, distinctive fossils or fossil groups are used to build up successions which can then be correlated either directly or in terms of general faunal resemblance. With a few exceptions rock-stratigraphical units are diachronous or time-transgressive, that is to say they are not of the same age throughout (figure 2) : in an advancing delta, for example, the foreset beds will be progressively younger seawards. In con-trast, biostratigraphical boundaries are taken to be approximately isochronous or time-parallel[9] because, for all practical purposes, individual faunal changes may be assumed to have occurred simul-taneously throughout the world.

How, then, do biostratigraphical and time-stratigraphical units differ? The American code concedes that, in practice, the geo-graphical extension of the latter will be influenced and generally controlled by lithology, fossil content and allied features. Many workers go further by arguing that time-stratigraphical boundaries can have no separate existence from the (largely biostratigraphical) criteria on which they are based. It is not surprising to find that certain units, namely system, series and stage, are common to chronostratigraphical schemes and to those that equate time-stratigraphical with biostratigraphical boundaries.

Homotaxis

In 1862, T. H. Huxley remarked that it would have been better for geology had so loose and ambiguous a term as 'contempor-aneous' been excluded from its vocabulary. *The Shorter Oxford English Dictionary (2 vols., 3rd edition)* gives as the second meaning of this word 'Of the same historical or geological period 1833', and cites Charles Lyell as the authority.[10] Huxley had been criticising with characteristic vigour the assumption prevalent among geol-ogists that similar fossil sequences in areas widely apart could be regarded as contemporaneous. He proposed *homotaxis* to signify similarity of 'serial relation' as distinct from *synchrony*.[11] It is widely conceded that Huxley had logic on his side; but (the argument runs) his remarks represented a retrograde step in the progress of stratigraphy because the procedure he was decrying, though wrong in its philosophical foundations, was essentially right in practice. Palaeontological correlation works because the time required for the dispersal of marine organisms (which are the chief concern of the stratigrapher) can safely be neglected when set against the vast spans of geological history.

Huxley bemoaned the fact that neither physical geology nor palaeontology were capable of demonstrating the 'absolute synchronism' of two strata.[12] The inability was not total: historical evidence had long been used to date recent geological changes in terms of calendar years, which meant that contemporaneous events could be recognised. But for prehistoric times there was no alternative to the use of fossils in correlation. We might speculate whether strict time-correlation would have been central to historical geology had geologists been empowered from the outset to assign dates to rocks. The fact remains that, through historical circumstances, it is not. 'Contemporaneous', 'simultaneous', and even 'synchronous' are taken to mean, not 'of the same age', but 'of the same geological age' (figure 3), with the implication that to all intents and purposes it often amounts to the same thing.

Age

Geological age is expressed in terms of the geological time scale.[13] This is an ordinal scale made up of units which are derived from the corresponding stratigraphical units and which are consequently unequal in length. A period, for example, represents the time during which a system was deposited; likewise an epoch corresponds to a series, and an age to a stage. These are listed in decreasing order of rank, but it does not follow that all periods, for example, are longer than all epochs.

The use of phrases such as 'quantitative time', 'absolute age', and 'chronological dating', and the existence of methods to determine the age of the earth and of rocks in years, suggest that the geologist has a second time scale at his disposal and that it is an interval scale. The ordinal time scale is based on one unidirectional process,

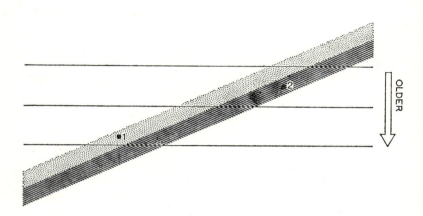

Fig. 3 Two versions of contemporaneity. The shaded strips represent steeply time-transgressive strata; the horizontal lines (time-lines) represent successive time increments. In accordance with the law of superposition, the lower of the two strata is taken to be the older; this holds good at any point along the contact between them, being a matter of relative age. But once rock-correlation is taken to connote time-correlation, geological equivalence risks being confused with contemporaneity: in the diagram, item 1, though stratigraphically younger than item 2, is chronometrically older.

that of organic evolution; the interval scale depends mainly on another, that of radioactive decay.

One writer has expressed the view that it is only a matter of convenience whether we adopt the 'relative' or the 'absolute' time-standard, as they are both partial standards derived from the fundamentally unknowable abstract or ideal time.[14] 'I have no patience', another author states, 'with the claim that organic evolution measures one kind of time and radioactive disintegration another.'[15] These are matters to be discussed later, and are mentioned here only to show that there are theoretical objections to

the distinction between ordinal and interval scales. More important
when it comes to practical geology is the consensus of opinion in
favour of palaeontology as the basis of geological time-reckoning.
The available radiometric dates are too few, too difficult to interpret or too unreliable; stratigraphers work mainly with marine
deposits, which lend themselves to palaeontological dating much
more than to radiometric methods; and in any case the fossil record
has the higher resolving power.[16] What trustworthy radiometric
dates there are must be reserved for calibrating the ordinal scale.
This is what generally emerges as the task of geochronology.

Quaternary stratigraphy

The Quaternary presents many obstacles to the application of
standard stratigraphical procedure. For example, much of the rock
record consists of thin, patchy and unconsolidated terrestrial deposits, even though the balance is being corrected by progress in
the study of ocean cores. Again, the time-spans at issue are generally
too short for pronounced evolutionary changes to have occurred,
while palaeontological correlation is further complicated by the
development of similar environments in different places at different
times as a result of the repeated climatic changes that characterised
the period. Yet, although some workers believe that the Quaternary
calls for techniques and criteria of its own,[17] procedural conformity
is the rule, notably as regards the retention of rock-stratigraphical
and time-stratigraphical units and the rules that govern their
correlation. Further, the stratigraphical units that have been introduced or developed to cope with the idiosyncracies of the Quaternary are conventional in design. The 'geologic-climate unit', for
example, is

> an inferred widespread climatic episode defined from a sub-
> division of Quaternary rocks. . . . At any single place the time
> boundaries of the geologic-climate unit are defined by the boun-
> daries of some kind of stratigraphic unit. These local strati-
> graphic boundaries may be isochronous surfaces, but the different
> stratigraphic boundaries that define the limits of the geologic-
> time unit in different latitudes are not likely to be isochronous.
> In this respect geologic-climate units differ from geologic-climate
> units, which are based on time-stratigraphic units.[18]

The definition as a whole, and even more the implication that time-stratigraphical units are time-parallel, has the authentic ring of stratigraphical parlance. Note that glacial-stratigraphical units may be used in preference to the above in order to free the term 'glaciation' from exclusively climatic interpretation : glaciation is not necessarily controlled by climate.[19]

In contrast, soil-stratigraphical units, which represent the alteration of one or more underlying rock-stratigraphical units by pedological (soil-forming) processes, are thought to be essentially time-parallel because they represent climatic interludes, which supposedly affected large areas, and perhaps even the entire globe, at the same time. Geopolarity units, a recent introduction, are time-parallel by definition, as they denote intervals characterised by normal or reversed polarity of the earth's magnetic field.[20]

Quaternary correlation

To judge from the last two units, true time-correlation would appear to be feasible in Quaternary stratigraphy. Nevertheless most workers stress the *approximate* time-parallelism of ancient soils, while those who champion their use in long-distance time-correlation rest their case on some hypothetical mechanism for producing world-wide climatic changes.[21] However extensive, volcanic ash beds are in global terms very localised, and even within one such area the simultaneity of ash deposition is strictly speaking not absolute. Raised beaches and other indicators of former positions of sea level would seem more dependable, since changes in ocean volume have presumably affected the entire globe simultaneously; their time-correlation, however, usually hinges on radiometric, palaeontological and allied techniques and therefore does not amount to a distinct contribution to the problem. Similarly, the contemporaneity of prehistoric remains cannot be demonstrated without external sources of dating.

The usual conclusion is that, as for earlier periods, reliance has to be placed on palaeontological methods supplemented by evidence from ash-falls, sea levels, ancient soils and other marker beds. Yet, as we have seen, the fossil record is both short and marred by the climatic vicissitudes that are a prominent feature of the Quaternary.

Paradoxically, these problems have helped palaeontology retain its habitual prominence. In 1849 Charles Lyell abandoned faunal definitions of the Pleistocene in favour of a climatic boundary. Climatic change has since come to dominate both the subject matter of Quaternary stratigraphy and the techniques it employs in dating and correlation. With a few exceptions, such as the mammalian faunas used in formulating broad subdivisions of the Quaternary, fossil sequences are matched principally in terms of the climatic successions they reflect expressed as glacials and inter-glacials or by some analogous scheme.[22] In a sense this is no depar-ture from pre-Quaternary stratigraphical practice, where fossil assemblages of restricted distribution may need to be correlated in environmental terms;[23] the difference is largely one of emphasis. The growing attention being paid to the oceanic record adds further to the role of palaeontology, although, once again, palaeo-climatic rather than evolutionary schemes are being most actively pursued especially where there is a need to correlate between land and sea.

Homotaxial reasoning thus enters Quaternary studies by two routes, the palaeontological and the palaeoclimatic,[24] as both lead to the assumption that similar sequences were approximately con-temporaneous. That these routes criss-cross follows from earlier paragraphs; it is manifest in, say, attempts to date deposits using pollen analysis to fit them into some standard climatic curve, in the correlation of archaeological horizons by reference to the environmental changes recorded at the sites, and in the dating of human fossil remains by relating their palaeoclimatic context to temperature curves derived from the isotopic analysis of deep-sea cores.[25] Climatic correlation has, in fact, been defended on the grounds that the boundaries it yields are not likely to be any more steeply time-transgressive than those based directly on palaeonto-logical criteria. As so few stratigraphical units are time-parallel in a narrow sense, 'one is obliged to live with time-transgression as the way of life of stratigraphy'.[26]

The above conclusion is qualified by a phrase to the effect that precise time-correlation is not yet possible between events for which radiometric dates are not available. As we have observed, such dates are essential for the time-correlation of sea levels, cultures and the like. Some palynologists readily concede that radiocarbon dating is increasingly taking over from pollen analysis the task of establishing the age of individual horizons;[27] an archaeologist has recently looked forward to the day when 'we can use artefacts, the basic material of archaeology, for purposes other than dating'.[28] Taken in conjunction with the improvements that are taking place in the reliability and scope of dating techniques, statements such as these might lead one to conclude that before long time-correlation will as a matter of course be based on an interval scale. But it is not merely a question of technology : the hard-won accuracy of the dates is put at risk once they are grafted onto time-transgressive stratigraphical units. Few stratigraphical tables for the Quaternary are now without calibration in years, centuries or millennia, but their subdivision is still governed by the usual criteria; similarly, dates obtained for archaeological horizons still tend to be extrapolated on cultural grounds. When correlation precedes dating it is difficult to argue about contemporaneity, let alone about age differences.

3 Chronometric age

. . . that duration of time which is marked, but not
measured, by the geological record.

D'Arcy Thompson, *On Growth and Form*

The triumphs of stratigraphy are in no way minimised by the
recognition of certain problems whose study is poorly served by
ordinal dating and correlation. Many of the problems are made
more tractable by the calibration of stratigraphical units with
radiometric and other dates; the remainder require for their solution a more drastic departure from the methods of William Smith.

To begin with, an ordinal time scale does not enable us to
measure time; it is therefore of little help in dealing with rates of
change. A. G. Fischer has recently proposed the adoption of a
standard measure of geological time–distance rates, the Bubnoff
unit, equivalent to 1 metre/1 million years.[1] Although the need
for such a unit can be questioned, the proposal has served to draw
attention to the wide range of phenomena amenable to analysis
in terms of both linear and time measurements. Most of the
examples Fischer gives bear on 'the physiology of our planet' (figure
4); the reference to physiology brings in its train the consideration
of forces, which in turn cannot be considered without reference
to the time dimension. Further, the inclusion of an entry for
skeletal growth is a reminder that the metering of cumulative
change is an everyday matter in the biological sciences and is
therefore germane to many aspects of environmental reconstruction.

Secondly, proposed time-correlations cannot be tested unless the
items in question are dated independently by reference to an
interval time scale. This was understandably one of the tasks to

Fig. 4   Mean rates of various incremental or time–distance processes plotted on a logarithmic scale in Bubnoff units. Slightly modified after Fischer, 1969, figure 1.

which radiocarbon dating was applied soon after its introduction.[2] Its logical extension is to use isochronous time-lines and time-planes as a reference against which continuous portions of the rock record (as opposed to isolated sections) can be matched in two and three dimensions respectively. The analysis of changes in facies, that is to say in the lithological and palaentological character of sedimentary units, is difficult unless one can show that the various environments coexisted; consider, for example, the wide range of facies that characterise a shoreline at any one time. Once contemporaneity has been demonstrated, rock type and fossil content can serve as environmental indicators. A similar approach may be fruitful in the study of archaelogical sites, in that artefact assemblages formerly regarded as representatives of different cultures may be reinterpreted as specialised tool kits employed by a single human group engaged in exploiting diverse resources.[3]

But it is in the search for mechanisms that the interval scale comes into its own. Many explanatory hypotheses hinge, at all events initially, on before-and-after relationships. A phase of faunal extinction cannot be convincingly ascribed to environmental changes that took place later; similarly, discussion of cultural diffusion is destined to remain barren unless the direction in which it operated is known.[4] The relative age of 'cause' and 'effect' (or donor and recipient) can be demonstrated by stratigraphical methods only where the law of superposition is applicable; elsewhere it will emerge only from their position on a common time scale. And once the processes themselves come under scrutiny, the need for an interval scale will be felt even where the relative age of the two items is self-evident.

Furthermore, some of the assumptions that are essential to chronostratigraphical correlation can be transformed by dating from implicit articles of faith into subjects for research. Critics of palaeontological time-correlation have long enjoyed pointing out that faunal similarity rules out contemporaneity simply because migration could not have been instantaneous. Its defenders argue that, to judge from modern parallels, the time involved is trivial : the rabbit for example, populated large areas of Australia soon after its introduction[5] (even if it did not reach Australia at all until it was introduced). Thus, the shortest time units that can be demarcated with the help of fossils within the Jurassic are held[6] to represent a degree of accuracy comparable with discriminating between an event in 1007 B.C. and one in 1000 B.C. A third group is content to remind us that, however successful the palaeontological subdivision of geological time, its theoretical basis must be borne in mind lest divisions which are permissible in practice be used in a quite inadmissible fundamental fashion.[7] The three viewpoints become redundant once it becomes possible to measure the rate at which a certain organism or assemblage did spread : all methods of correlation by 'signal',[8] including the use of volcanic ash, demand information on both the direction of transit and the velocity of the signal. Later sections will try to show that debates over the validity of climatic correlation can also be rendered superfluous by dating.

Time-measurement

A dip into the literature devoted to the philosophy of time
provides reassurance that geology, which was prominent in sowing
doubts regarding the validity of biblical chronology,[9] is not alone
in finding the harvest something of a handful. Besides *Schaden-
freude,* the dip (and a plunge would be ill-advised) provides the
basis of an approach to time-measurement which appears to satisfy
the demands both of logic and of expediency.

The logical problems facing time-measurement remain intract-
able unless we make a preliminary coordinative definition, namely
a definition of a unit of time. If this is to serve for measurement
we need a second metrical coordinative definition, one which
makes successive time-intervals equal.[10] In choosing both the unit
and the clock with which to define it we are influenced by the
need for descriptive simplicity : we seek a time scale which simplifies
the description of nature.[11]

Clocks which depend on the gravitational constant appear to
measure the same time as those that are governed by nuclear
constants.[12] The first group includes the 'earth clock' or, more
specifically, the periodic process consisting of the earth's orbit
around the sun, as this is a gravitational motion whereas the earth's
rotation is an inertial motion. The second group includes atomic
clocks. The earth clock is of course widely applicable, and there are
strong grounds for assuming that the periods it marks off have
remained constant. The fact that atomic clocks form the basis of
international definitions of time presents little problem, as the year
(or a particular year) can be calibrated by reference to such a
standard. In 1956 the second was defined as a specific fraction of
the year 1900. In 1966 it was redefined in terms of the frequency
of radiation of the caesium-133 atom between specified energy
levels, and further redefinitions may be expected as the quality of
atomic clocks is improved.

The year also has the virtue of familiarity. It has been opposed
on the grounds that it is redolent of anthropocentric thinking :
'there is no fundamental reason—other than habit—why an increment of this exact length should be used. Any scale based on *any*
equal increment of time will serve as well.[13] But if any unit will

do as well as the year, so will the year do as well as any other unit. As it is, the very act of choosing any unit could be said to confer anthropocentricity upon it.

The above procedure yields a time scale whose application to earth history raises problems which, though considerable, are purely technical. The need remains to give it a name. Oakley proposed the adoption of 'chronometric dating' to denote the actual age of a specimen, and the use of 'absolute dating' to convey position on a time-line without an age in years being specified : thus two deposits known to be contemporaneous (say because they both contain a certain volcanic ash) could be said to have the same absolute age.[14] Holmes, who thought 'absolute age' a meaningless term on the grounds that age did not become absolute through being expressed in units of time such as the year, preferred to speak of 'radiometric age' or simply 'age'.[15] Others have suggested that 'chronologic' be used for the time scale derived from radiometric methods and 'geochronologic' for the geological time scale.[16] The simplest solution would be to use 'age' unless there is risk of confusion with the concept of geological age, and 'chronometric age' when the issue of time-measurement needs to be stressed. Chronometric age is, of course, derived from methods other than those based on radioactive decay.

Simultaneity

Further judicious reference to the philosophical literature leads to the definition of simultaneity as equality of time values on parallel times scales, a coordinative definition (our third) which enables us to complete the logical structure of time-measurement. As Reichenbach points out, the definition may be tautological, but so are all conceptual definitions because they are concerned with analytical relationships.[17]

Time-correlation now emerges as nothing more than the corollary of chronometric dating. Events which have the same apparent age are accepted as being contemporaneous, whereas those that are not can be linked in terms of a before-and-after relation. In neither case is the bond indissoluble, since all measured ages are approximations open to correction and refinement, rather as spot heights on a map will need to be revised to allow for advances in the techniques of surveying. It has been argued that basing a stratigraphical system on the notion of simultaneity imparts to it a high degree of instability, simultaneity being a relationship that is difficult to confirm and easy to falsify.[18] The objection disappears once simultaneity is disengaged from stratigraphical definition and comes to serve as part of the framework on which time relationships are erected.

This approach is already followed for a variety of purposes. In hydrological studies, for instance, the general trend of storm movement can be established with the help of isochronal maps on which the isolines (lines of equal value) represent successive 'time contours' for a particular event, such as the onset of precipitation. Isochrones have also been employed on maps depicting the progress of deglaciation in North America (figure 5).[19] ('Isochron' is a more euphonious rendering but it has been pre-empted for certain diagrams used in the analysis of radioactive decay systems.) As on a topographical map, some interpolation will be required however

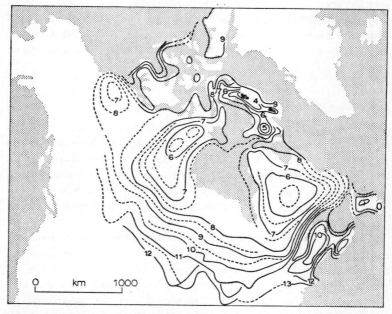

Fig. 5 Radiocarbon isochrones marking the progressive retreat of the Laurentide ice sheet. Modified after Bryson *et al.*, 1969, figure 1. Figures are multiples of 10^3 radiocarbon years.

dense the network of point measurements, since the isochrones will need to be separated at an arbitrary interval. On the storm map the spacing may represent two-hourly periods, on the deglaciation map it may amount to increments of 500 years. Other units can be used provided they are readily expressed (if need be) as fractions or multiples of the year.

Newton recognised two kinds of time, the one 'absolute, true, and mathematical', and the other 'relative, apparent and common'. The former conforms with the physical time of some philosophers and with the definition adopted here in that the units furnished by the earth clock or by atomic clocks are empirically in accord with natural processes. In brief, it forms part and parcel of our acceptance of uniform laws of nature. The second brand of time, sometimes referred to as experienced, concrete, subjective and historical time,[20] can be dispensed with; as we have seen, it is experience (though not all kinds of experience) that forms the basis of our choice of time scale. Moreover, the concept of historical or concrete time leads to a fragmented view of earth history. The age of reptiles prevailed even in those areas where reptile bones are absent from the fossil record.[21]

Dating methods

The biological time scale used by the historical geologist is related to the interval scale as a pedigree might be to a calendar. As things now stand, 'Rates of evolution do not measure time. Time, on the contrary, measures rates of evolution',[22] whether expressed in somatic features or in the behavioural changes that are represented in the archaeological record.

Yet it is unwarranted to dismiss the possibility that an evolutionary clock may ultimately be developed, witness the remarkably successful attempts by Lyell and others to estimate the duration of geological periods by reference to evolutionary 'cycles' and 'species-steps'. It is no less rash to assert[23] that neither theoretical nor empirical grounds exist for supposing that any historical sequence of biological events might mark off uniform intervals of time. One need only consider the remarkable stability of circadian rhythms to picture some more permanent (that is, susceptible to fossilisation) echo of astronomical cycles and hence of periodic events that fall within reach of correction. As it happens, the growth bands of certain fossil corals appear to bear the imprint of daily, monthly and yearly cycles, and the deviations in the rate of rotation of the earth which they indicate tally with those obtained from theoretical calculations.[24]

These corals are too localised to yield a continuous time-sequence, and their age is obtained from other sources. But astronomical periodicities already underlie several dating techniques which have found wide application in Quaternary studies, namely dendro-chronology, the analysis of varves and other rhythmic sediments, and the many variants of the solar radiation method. At first glance they are unexceptionable. In favourable circumstances tree-rings and varves will reflect annual cycles, the former of growth, the latter of deposition, and by counting successive layers it will be possible to build up a chronology which is axiomatically expressed in years. Similarly, if we accept that periodic perturbations of the earth's orbit around the sun have produced fluctuations in the distribution of solar radiation over the surface of the earth, we can date associated geological features (such as glacial phases) directly.[25] The finding that particular years may be represented by two growth rings or no rings at all, that some varves may be diurnal or the product of storms,[26] and that many of the orbital perturbations appear to have had no geological echo, can then be accommodated in our category of 'technical problems'.

Yet, despite the achievements of dendrochronology and varve analysis in the south-western United States, Scandinavia and else-where, they are techniques which do not meet the strict require-ments of chronometric dating. As with coral bands, the links between the astronomical periodicities and their terrestrial mani-festations are indirect, complex and hypothetical. The final arbiter in judging the validity of 'rhythmic' chronologies remains an independent dating source, usually historical or radiometric; and if there is agreement, it deserves the puzzled respect we accord to Lyell's estimates of the duration of the Tertiary. Where tree-rings are concerned such caution must seem absurdly exaggerated; but it is legitimate in an attempt to differentiate between validity and success. (The fact that these methods are not widely applicable, and hence thrive on stratigraphical correlation, is of course an important practical consideration.)

In contrast, it is inherent in the nature of radiometric methods to yield dates which, like those of the historian, can in principle be mapped directly on to the chronometric scale. The law of radioactive decay takes the form

$$\mathrm{d}n/\mathrm{d}t = -\lambda n$$

where $\mathrm{d}n/\mathrm{d}t$ is the number of atoms of the radioactive isotope decaying per unit time, n is the number present, and λ is the decay constant. (The negative sign indicates that n is decreasing.) Integration gives

$$n = n_0 e^{-\lambda t}$$

where n_0 is the number of radioactive atoms present initially (that is, when $t = 0$). This equation can be solved for t. The decay constants on which the methods depend express the probability that a given radioactive nucleus will decay in a given time interval. There is every indication that they are immutable in any terrestrial environment; that they have remained unchanged in a conclusion implicit in our acceptance of universal laws of nature and sanctioned by the data at our disposal.[27]

Uncertainties regarding these constants, calibration standards, and other values entering the calculation, like errors of measurement, do not undermine the validity of radiometric dating even if they clearly influence the accuracy of the resulting ages, that is to say the extent to which these deviate from the true (if unascertainable) ages. The same applies to the statistical errors stemming from the random nature of radioactive decay, and to the other factors that affect analytical precision, that is to say the reproducibility of the determination.

It is misleading to group radiometric dating techniques together with those that depend on sediment thickness, the extent of erosion or weathering, and the ocean's salinity, under the collective heading of 'hour-glass methods'. Hour-glasses are of little value unless the sand is known to flow at a uniform rate. The analogy is therefore erroneous as regards radiometric methods because decay curves are logarithmic and not linear; but this is not a serious issue. Much more important is the fact that rates of weathering, erosion and deposition cannot be assumed to have remained constant even if— as in the case of palaeontological dating—some early estimates based on this assumption gave results that come close to values currently in favour. It follows that the use of sediment thickness as a means of bridging the gaps between the available radiometric dates[28] can be condoned only as a temporary measure.

There is little need to point out that both archaeological dating and reference to reversals of the earth's magnetic field are methods of extrapolating radiometric ages rather than sources of dates in their own right. Artefacts found within a deposit can be matched with similar material from stratified sites for which dates are to hand. Quite apart from the fact that they are often derived and hence indicate only the maximum age of the deposit, their correlation with the dated material involves homotaxial reasoning and, if practised over long distances, can lead to serious error. Reversals of the earth's polarity were presumably simultaneous over the globe and are therefore free from this taint; but where (as is generally the case) geopolarity sequences are correlated in purely ordinal terms, radiometric dating is still required to check the proposed correspondences.

Dating the last 20 000 years

Let us now turn to the period which is to supply most of the case-studies. Of the radiometric methods that are relevant, radiocarbon (^{14}C) dating is the most generally useful in being applicable to a wide range of organic materials. Furthermore, as the effective range of the method goes back beyond 40 000 years ago, both the period under review and the events that led up to it can be dated without the problems raised by the use of two or more incompatible decay schemes.

The errors that arise in its application vary greatly in their significance and intractability. For example, little can be done about the problem of statistical uncertainty beyond extending the period of counting, but improvements in laboratory technique have improved matters over the last few years. New determinations of the half-life of radiocarbon, that is to say the time required for half a given quantity of radioactive atoms to decay, can easily be allowed for. The value that is at present in use internationally is 5568 ± 30 years (the plus-or-minus figure representing, as in radiocarbon ages, the 'statistical uncertainty' to one standard deviation). Were a more recent determination giving 5730 ± 40 to be adopted, existing dates could be increased by the requisite 3 per cent. Again, errors due to the fact that different datable materials incorporate carbon isotopes in different proportions can be assessed by making a separate $^{13}C/^{12}C$ determination.

The situation appears to become more serious once we turn from the laboratory to the field. For example, the inclusion of 'old' carbon from groundwater flowing out of a limestone may give ages higher than they should be; conversely, 'young' dates may result from contamination of the sample by living organisms. Nevertheless a close study of the site being sampled will at least enable the likely sources of error to be recognised so that suitable corrective measures can be attempted by the analyst. In some circumstances the use of comparative material makes it possible to gauge the magnitude of the distortion,[29] or at least to specify the probable range of values. Thus the age of an alluvial deposit in Mexico was taken to lie somewhere between 15 550 B.C. (a date obtained from tufa, and therefore likely to be excessive) and 8000 B.C. (determined from charcoal which showed evidence of contamination by younger material).[30] If the worst comes to the worst, a suspect date can always serve as a maximum or minimum, whether because no such assessments are feasible or because it is not known how long after organic death the sample was included within the section. The fact remains that contamination by young material has little effect on dates for the last 20 000 years, and that, as regards continental carbonates susceptible to distortion in the opposite direction, 'if enough information is available to make suitable corrections, accuracy and precision may be closely defined'.[31]

The above arguments may appear to protest too much; yet the loss of nerve displayed by some workers confronted by the possibility of error is no more commendable. The phenomenon has been particularly in evidence following the finding that radiocarbon dates are in disagreement with ages derived from historical evidence, tree-ring counting and varves (figure 6), and has led, if not to the abandonment of radiocarbon dating, at least to the publication of ages cautiously expressed in 'radiocarbon years'. Granted the validity of the 'known' ages,[32] it would seem simpler to make the required correction to new and existing determinations in the knowledge that the correction—like all the other values that enter into

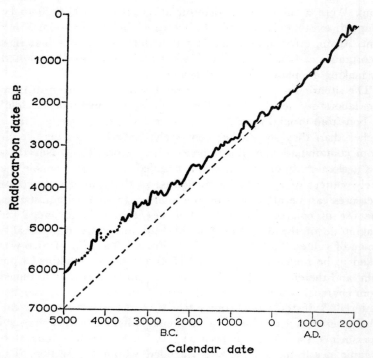

Fig. 6 Apparent divergence between radiocarbon and calendar years, based on tree-ring measurements on Bristlecone pine (*Pinus aristata*). Modified after Suess, 1970.

the calculation—is likely to witness further refinement. (The corollary is of course that all the steps leading to a provisional date need to be spelled out if only in coded form.) And, as our understanding of the reasons for the discrepancies improves, so will it be possible to make the correction for dates beyond the reach of history, tree-rings and varves, just as the distortion produced by the burning of fossil fuels and the testing of atomic weapons can be allowed for without having to depend on dubious empirical data. As regards the 'non-uniqueness' of some of the resulting dates[33]—in that calibration by means of a graph such as that given in figure 6 may mean that the radiocarbon method is unable to distinguish between, say, wood that grew in A.D. 1690 and wood that grew in A.D. 1850—a series of closely-spaced determinations spanning the crucial portion of the sequence may permit some or all of the problematic readings to be eliminated by reference to their relative vertical position.[34] To sum up :

> All radiocarbon dates should be viewed critically, using information provided by the original report and making reasonable corrections and realistic estimates of uncertainty. The age should then be applied to a problem only to the extent that it is truly useful.[35]

Other radiometric methods at present have little application to the record of the last 20 000 years, but the situation may be expected to change as they are further refined and modified. The protactinium method (^{231}Pa) and the protactinium–ionium method (^{231}Pa/^{230}Th), when applied to various materials, give dates that often agree well with those obtained with radiocarbon. The ionium deficiency method (^{230}Th) can on occasion be used in dating organic carbonates less than 5000 years old (figure 7)[36]; potassium–argon (K/Ar) dating of rocks as little as 10 000 years old is technically feasible, the expected standard deviation of precision being in the region of $\pm$ 20 per cent.[37] Fission-track dating can be used for crystalline or glassy material able to retain the tracks and containing sufficient of the uranium-238 whose spontaneous fission is responsible for the damage trails, including obsidian and some man-made glasses. In terms of the arguments advanced earlier, all these techniques are preferable to others which involve assumptions about the rate at which variable processes operate, among them obsidian dating by reference to the depth to which hydration has penetrated,[38] the use of lichens in formerly glaciated areas to determine the time that has elapsed since the rock surface was last disturbed (lichenometry),[39] and the ion exchange method of dating alluvial deposits.[40] What the position is regarding the dating of pottery by thermoluminescence remains to be seen.

As we approach the immediate past, historical evidence begins to supplant radiocarbon as the major source of dates capable of expression on the chronometric scale. This is hardly a novelty : sherds, ruins, verbal accounts and popular traditions were fully exploited not only by Lyell and von Hoff but also by Renaissance scholars and the Greeks and Romans before them. Yet they subsequently lapsed into relative neglect, largely because the triumph

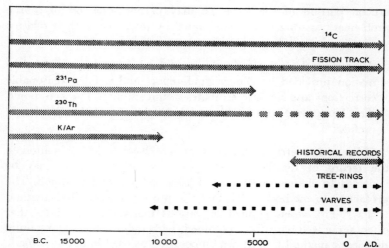

Fig. 7   Approximate range of various dating techniques within the last 20 000 years. The dashed portion of the ^{230}Th line indicates the additional period spanned by the ionium deficiency method. The tree-ring and varve lines are dotted to represent doubts raised in the text.

of uniformitarianism (*see below*, p. 108), to which such evidence was a major contributor, diverted the attention of geologists to the wider realms of remoter earth history.

Besides supplying ages, historical sources serve the wider function of enabling models put forward to account for changes in the period of the earth's rotation to be tested directly (though by no means conclusively). Geological methods span much longer intervals but are circumscribed by their dependence on rhythmic deposits, the growth rings of corals, and similar supposedly periodic phenomena. Admittedly the historian is no less prone than the geologist to wishful correlation and a yearning for periodicities; in the present content, the kind of history that gives more weight to monarchs and battles than to social conditions makes up in utility what it lacks in sparkle.

The extension of the year-based time scale into the historical period raises a minor problem, that of the B.C./A.D. divide. Some historians, not all of them Marxist, use (−) in place of B.C. and (+) for A.D., a solution which copes with agnostic scruples and also with the comical effect of a date such as 458 500 B.C. One alternative is to apply the B.P. (before present) notation, which is already the rule in radiocarbon dating, to all dates however they may have been derived; but as the 'present' has been defined as the year 1950, later events must be deemed to lie in the future, a trivial issue where millennia are under discussion but of concern to the historian. Some workers accordingly compromise by using B.P. for Pleistocene ages and the B.C./A.D. system for dates falling within the last 10 000 years. Others favour using B.P. for uncorrected radiocarbon dates and the B.C./A.D. system for radiocarbon dates that have been corrected in the light of new half-life determinations, tree-ring chronologies and the like. Another view is that historical chronology and radiocarbon chronology should not be mixed in view of the totally different character of the data on which they are based. The prehistorian is accordingly advised to go on using the conventional radiometric time scale.[41]

The solution adopted in this book is to use B.C./A.D. dates unless the point to be stressed is the total time that has elapsed since the event in question took place, whereupon 'years ago' will be used in preference to B.P. Where broad subdivisions or approximations are needed, the power of ten notation (10^3 years, etc.) has been found preferable to terms such as 'decachiliad' or 'megacentury'.

Dates and events

We have already considered the problems of analytical accuracy and precision. Let us conclude with the attribution of dates.

Many age determinations can be related to stratigraphical units without difficulty, if not always as intimately as might be desired. Ash-falls and lavas commonly yield dates which refer to the time of their accumulation. Archaeological and radiocarbon dates which depart from the ideal can still provide limiting ages for the deposit or erosional feature. Metamorphic rocks may give discordant ages when subjected to different radiometric methods, but the time of their emplacement can often be inferred from the date given by the most resistant parent–daughter system analysed or by the application of whole-rock assays.

But there are occasions when the dates defy the efforts made to give them stratigraphical meaning. Some of the ages obtained for a metamorphic rock may refer to episodes of heating or crystallisation that followed emplacement. The provenance of dated artefacts or fossils may be uncertain. Or, worse still, the dates will clash with what the stratigraphy had led the investigator to expect.

At this juncture it is useful to recall that geochronologists sometimes use the term 'event' to distinguish 'whatever it is that happened from more specific geologic phenomena'.[42] Viewed in this light, discordant ages which tell us little about emplacement can prove informative about geochemical history;[43] for instance age provinces which represent particular ranges of radiometric

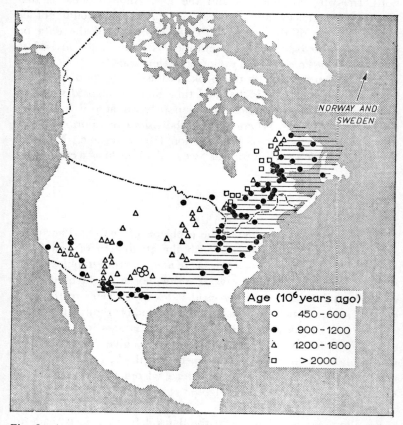

Fig. 8   Age province based on radiometric dates of 900–1200 million years ago. After Wasserburg, 1966, figure 3.

dates (figure 8) may help to identify thermotectonic zones. In brief, there is little cause to fear that, if dates cannot be fitted into the classic stratigraphical mould, 'the tail may well wag the dog' :[44] there is nothing inherently commendable in the stratigraphical dog wagging the dating tail. By accepting the possibility that not all dated events lend themselves to stratigraphical translation we safeguard the integrity of the age determinations possibly at the price of having to shelve them.

The late Quaternary is sufficiently rich in well-documented dates for such prodigality to be tolerated : a recent estimate put its gain in published radiocarbon dates between 1959 and 1969 at 10 000. Where stratigraphical attribution is straightforward nothing is involved beyond the refusal to treat the date as an elastic capable of being stretched wherever the stratigraphy may lead. But where the attribution is debatable it is the date which must be given the benefit of the doubt.

The time may not yet be ripe for the distant past to be treated in similar manner; yet appeals to the unripeness of time[45] are not always founded on hard fact. As it is, some workers already feel the urge to exploit whatever radiometric dates there are on the grounds that failure to do so 'results in unnecessarily erroneous correlations and silently perpetuates the outdated "equal space—equal time" concept'.[46] As was noted in the *Preface*, radiometric methods already play a prominent part in Precambrian studies; even if this is a case of *faute de mieux* something appears to have been gained from not having to compromise with an existing palaeontological framework. It is therefore all the more regrettable that recent fossil finds should have stimulated renewed interest in the 'time-honoured' methods of palaeontology, and that glaciation should be mooted as the possible basis for subdividing the Precambrian record.

4 Subdividing the record

What's in a name?
Romeo and Juliet

In 1924 no less than in 1862 (when Huxley was delivering his strictures on homotaxial correlation), it was reasonable to conclude that

> while we cannot now, and probably may never, hope to divide geologic time into centuries and millenniums, we can divide it into periods, each of which has its own special significance in the history of the earth.[1]

Circumstances have changed; the periods have endured.

There are clear advantages in having names to denote broad subdivisions of geological history, and in adhering to names that are generally accepted rather than developing local equivalents. The danger lies in allowing nomenclature to gain the upper hand. 'The tyranny of names', whose influence on Precambrian studies was deplored by Holmes,[2] can take many forms. At its most extreme it perpetuates units which are no longer sanctioned by the evidence or by the benefits they bring. It may operate more deviously by ascribing to a particular unit more than was intended when the unit was first defined. And it can promote the refinement of boundaries whose chief virtue had been their elasticity: the fact that we can now divide geological time into centuries and millennia has sometimes ossified rather than outmoded existing subdivisions.

The issues of definition and delimitation are prominent in Quaternary studies, witness the perennial debate over the nature and position of the Pliocene–Pleistocene boundary. This chapter is chiefly concerned with the Holocene because it is a unit which is especially prone to reassessment in the light of radiometric dating.

The Holocene

The name Holocene (from the Greek ὅλος whole, entire, and καινός recent) was introduced by Gervais in 1869; in 1885 the Portuguese Commission at the International Geological Congress suggested that it should rank as a Tertiary 'stage' to follow the Pleistocene. Since then Holocene has come to be widely used, particularly in Europe, as the second of the epochs into which the Quaternary period is divided (table 1). Its American equivalent, Recent, has a longer and more confused history. In 1833 Lyell proposed the name of Recent epoch for the whole of post-Pliocene time; in 1839 he subdivided this into post-Pliocene and Recent, and introduced Pleistocene in the place of newer (or New) Pliocene. In 1846 Edward Forbes suggested using Pleistocene for post-Pliocene and equated it with the Glacial epoch that had recently been recognised. Lyell's initial reaction was to 'abstain' from the name in order to prevent confusion,[3] but in 1873 he decided to

TABLE 1.

Geological time units	Era	Period	Epoch
Stratigraphical units		System	Series
	Cainozoic	Quaternary	Holocene
			Pleistocene
		Tertiary	Pliocene Miocene Oligocene Eocene Palaeocene

adopt Forbes' proposal. The Recent was thus identified with post-glacial times and in 1903 it was officially recognised by the United States Geological Survey as the latter part of the Quaternary period, only to be supplanted by Holocene in 1968. It is worth adding that Quaternary began life in 1829 as the French equivalent of post-Pliocene, and that, although it is generally ranked as a period, some authors elevate it to an era while others prefer to do without it.

The synonyms that sometimes replace Holocene or Recent include post-Würm, Postglacial and, for lower latitudes, Postpluvial. *Attuale* is used by some Italian geologists, as 'actual period' was by Lyell when first defining the Recent.[4] Further examples are Late-glacial, Technogen, Neotechnical epoch, Neothermal, late or upper Quaternary subdivision, post-Valday time and Kainochron.[5] There are doubtless others.

Holocene and Recent serve both for the epoch and for the time-stratigraphical unit corresponding to it, namely, the series. We also meet pairs of terms, notably, the German *Alluvium* for the series of the *Postglazialzeit* epoch, and *Diluvium*—a Catastrophist hang-over—for the series of the *Eiszeit* or *Glazialzeit* epoch. Further, the marine Holocene deposits of Europe sometimes go by the name of Flandrian.

The continued use of more than one of the above alternatives is inconvenient, as when relevant material in a book is indexed partly under one name and partly under another;[6] more important, it adds unnecessarily to the existing terminological Babel. Unfor-tunately none of the names appears to meet with general approval, though usually for good reasons. *Attuale* means 'current or present', and strictly the use of post-Würm brings post-Wisconsin, post-Weichsel and other equivalents in its train. Postglacial, Postpluvial and the like are both too specific and not specific enough : glacials and pluvials are regional rather than global phenomena and they have occurred more than once. As it happens, Postglacial was already giving trouble in the late nineteenth century, when some geologists applied it to beds intermediate between the Glacial and the Recent and others found it was 'open to great objection on etymological grounds'.[7]

Etymology has more recently been invoked to demonstrate that the use of Recent is illogical as Pleistocene is derived from the Greek for 'most recent'.[8] From this point of view Holocene is even less acceptable.[9] Yet a Pandora's box would be opened by a rigorous analysis of geo-Greek. To all but the classicist, or the Hellenic and possibly the Rumanian geologist, the dead languages have the virtue of general incomprehensibility even though the first formulation of a neologism based on Greek or Latin roots clearly merits serious thought. On this count, at any rate, Holocene appears preferable to Recent which, even when capitalised and preceded by the definite article, remains patently confusing. In the same vein, Lyell had criticised the use of 'contemporaneous' for Recent because, 'as the word is so frequently in use to express the synchronous origin of distinct formations, it would be a source of great inconvenience and ambiguity, if we were to attach to it a technical sense'.[10] Holocene is furthermore consonant with other Cainozoic epoch names, such as Pliocene and Palaeocene,[11] and it is also well fitted to an international role in not inviting translation (as opposed to minor orthographic modification into the local vernacular). Recent could thus revert to informal usage, as Flint suggests albeit in his case so as to dispose of the epoch entirely.[12]

One argument against retaining a Holocene unit is that the Pleistocene is still in progress. Another is that the period it represents (figure 1) is too brief to merit epoch status, and a third that the record it has left is too patchy to make the Holocene worth recognising. Even if retention is favoured there is argument over its rank. R. G. West points out that the boundary between Holocene and Pleistocene does not differ from those between earlier interglacials and glacials. Hence the Holocene is better viewed as the most recent stage of the Pleistocene rather than as a separate series, and should be renamed Flandrian in accordance with the custom in north-west Europe of naming temperate interglacial stages after the marine transgression that characterised each of them.[13]

These threats in turn lead to defensive moves. It is claimed, for example, that the 'operational convenience' of a Recent epoch fortifies 'the already potent argument' against modifying any system of classification unless it is shown to be actually illogical.[14] J. K. Charlesworth made a similar plea for the Quaternary, namely that a term can be retained 'for its convenience if not for its scientific value'.[15] One can thus argue that the Holocene merits the rank of epoch by virtue of the wealth of stratigraphical detail it has already yielded, geological time-subdivision being largely a matter of expediency : duration 'has never been a determining factor in defining time units'.[16] Be that as it may, the abolitionists are unlikely to prevail in the face of a Commission of the International Association for Quaternary Research (INQUA) entirely devoted to the Holocene.

The Pleistocene–Holocene boundary

According to R. B. Morrison, Quaternary geologists in the United States have the same attitude towards this boundary as towards morality : they agree that a line has to be drawn somewhere, but they have been unable to agree on where it should be drawn.[17] The same may be said of the Pliocene–Pleistocene boundary, of concern even to those who reject the Holocene.

When Lyell first introduced the name Recent he used it to denote the period during which 'the earth has been tenanted by man'. In 1863 he redefined it to embrace 'those deposits in which not only all the shells but all the fossil mammalia are of living species', in accordance with the procedure he had adopted simultaneously with Desnoyers (author of the name Quaternary) of subdividing the Tertiary strata 'according to the different degrees of affinity which their fossil testacea bore to the living fauna'.[18] Once the Pleistocene had come to connote glacial conditions, however, the Recent was increasingly defined in climatic terms. Faunas and floras continued to play an important part but this time as environmental indicators rather than as the products of evolutionary change. A similar change of emphasis typified the use of fossils in delimiting the base of the Pleistocene.

Lyell foresaw trouble. 'Cases will occur where it may be scarcely possible to draw the line of demarcation between the Newer Pliocene and post-Pliocene, or between the latter and the recent deposits; and we must expect these difficulties to increase rather than diminish with every advance in our knowledge . . .'[19] In fact, demarcation is often easier than it was in Lyell's day; the problem lies more in making a choice from the numerous rival definitions that are now available. It may be noted in passing that the Recent of 1863 is not the Recent of modern workers. Lyell himself had witnessed the application of Eocene, Miocene and Pliocene to items which were not considered in his original definition, and he reminded his colleagues that the terms had been 'originally invented with reference to conchological data'. In the light of their abuse it is a moot point whether Lyell would concur with the claim that the identification of the three epochs represents his most important contribution to geology.[20]

Climatic definitions

The physical indicators that have served to mark the onset of postglacial conditions include sea level, fluctuations of the ice margin, the isotopic composition of planktonic Foraminifera, deltaic deposition in glacial lakes, waterfall retreat and the development of weathering horizons. The organic evidence includes pollen and other plant remains, terrestrial and marine faunas, and traces of human activity. It is unreasonable to expect any two indicators to yield the same boundary even if they refer to the same locality, especially where one of the indicators depends on glacial melting and is consequently subject to the retardation introduced by the latent heat of fusion of ice. Fortunately most local definitions of the Pleistocene–Holocene boundary hinge on a single palaeoclimatic source, either for the sake of simplicity or because no other evidence is readily to hand.

Where the crucial event is dated by 'hour-glass' methods, the result cannot be treated with any confidence. To take an extreme example, estimates for the time required for the retreat of Niagara Falls, which is one such measure of postglacial time, range from 7000 to 39 000 years.[21] The application of radiocarbon dating, whether directly or in conjunction with varve counting, has eliminated this problem; the choice of a diagnostic event has thereby gained the prominence it deserves.

The obvious candidate would seem to be glacial retreat, but it is increasingly clear that different parts of the ice margin retreated at different times and at different rates, and that readvances were commonplace; hence the search for episodes which marked the onset of decisive deglaciation. The bipartition of the Swedish ice at Ragunda is one such episode which deserves a place in the annals of Quaternary geology through its association with the pioneer work on varves by G. de Geer, but its diagnostic value is low since all we can say of events following bipartition is that 'the remaining ice probably disappeared rather quickly'.[22] Events as arbitrary as this are unlikely to meet with general approval.

Other definitions relate to climatic oscillations which were supposedly felt over large portions of the globe. A century ago T. M. Reade termed *recent* certain deposits which had been laid down during the last episode of land submergence.[23] Granted the acceptance of glacial melting rather than land subsidence as the driving force, this submergence—the Flandrian transgression—seems acceptable in being world-wide in its incidence and in having left distinct erosional and depositional traces.[24] Nevertheless Reade's definition includes within the Recent the latter part of conventional glacial times, as it may be supposed that sea level was at its lowest during the glacial maximum. The alternative is to select on the sea-level curve a point which indicates the corresponding glacial minimum. Some authors refer to a flattening in the curve 5000–7000 years ago, on the assumption that this was when the continental ice sheets had assumed their present relatively stable volumes.[25]

Two important objections to any sea-level definition are that it is difficult to recognise inland and that the climatic changes it betrays may have been restricted to the glaciated territory. More generally applicable (in both respects) are definitions based on pollen and fossil soils on land, and on palaeotemperature indices at sea. Pollen is widely favoured because it is an unambiguous and sensitive indicator which can be applied to most floristic zones. The changes in climate revealed by pollen analysis, especially as regards temperature, can often be traced over long distances, and radiocarbon dating shows that some of the boundaries between pollen zones are almost time-parallel. Thus, according to one palynologist 'The upper boundary of the Late-glacial is determined by the definite climatic improvement of the Holocene following the Younger Dryas pollen zone III. By means of pollen analysis this time may be accurately defined palaeontologically, the year being 8300 B.C. . . . and is universally accepted'.[26] In some areas weathering profiles or geosols have been found a useful complement to pollen analysis. Indeed, their advocates claim that geosols are more widely distributed that good pollen profiles and often remarkably time-parallel. A group of geosols whose development ended about

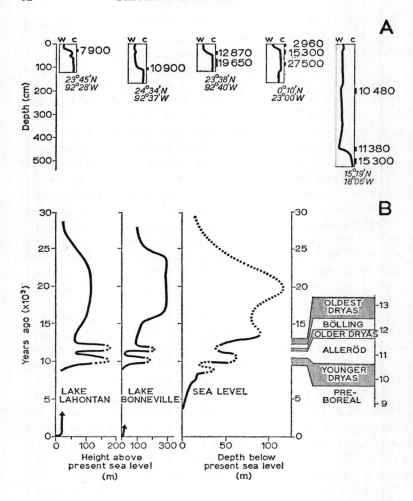

Fig. 9 Dating the Pleistocene–Holocene boundary. (A), Atlantic surface-water temperature curves based on Foraminifera (w = warmer, c = cooler) and ¹⁴C dates for selected deep-sea cores. (Error values for ¹⁴C dates have been omitted.) Modified after Broecker et al., 1960, figure 2. (B), Fluctuations in the level of two lakes in the Great Basin and in sea level, and European pollen zones corresponding to the transition between glacial and postglacial times. Modified after Curray, 1961, figure 1, based on various sources.

12 000 years ago is accordingly proposed as the Pleistocene–Holocene boundary for large parts of the United States.

Similarly, evidence has been obtained from ocean cores for an abrupt world-wide warming about 11 000 years ago, with radiocarbon ages for the event allegedly falling within 1000 years of that figure. It would appear that there is a close correspondence between the trends indicated by the relative abundance of certain species of planktonic Foraminifera, the oxygen-isotope curve, the onset of stagnant-water conditions in the Cariaco Trench off northwestern Venezuela, and a pronounced rise in the sedimentation rate of both carbonates and clays in the mid-equatorial Atlantic. These items are also held to match various continental indicators (figure 9).[27]

Such confident claims are balanced by less favourable verdicts. A recent review of Pleistocene boundaries drawn by means of pollen analysis speaks of 'the mess we are in at present' and emphasises the disparities in age between the various regional proposals.[28] In addition there are large areas where the survival of pollen and soils was not favoured, or where the available material awaits investigation. The latter point also applies to large parts of the ocean floor and militates against the drawing of global boundaries. Moreover the existing ocean-core evidence has been reinterpreted to show that the temperature rise at the close of glacial times, rather than being abrupt, spanned a period of some 7500 years;[29] while the distortion of the isotopic 'palaeotemperature' record by the introduction of glacial meltwater renders this facet of the marine record (like sea-level history) less representative of oceanic than of continental conditions.

The above are some of the matters that arise in considering the eligibility of existing boundary proposals. One general point which must take precedence over further examples is the emphasis placed on time-parallelism both in championing and in challenging the criteria used and the boundaries they yield. Even where the case is argued for a definition which conforms to accepted usage in depending on time-stratigraphical units, the indicators that are favoured tend to be those which come closest to time-parallelism. This ideal is embodied in proposals for an international stratotype (or reference section) to which provincial stratotypes would be linked with the help of radiocarbon dating.[30] The same ideal serves

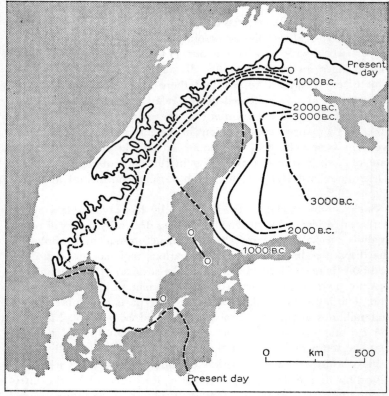

Fig. 10 Diachronism of a climatic boundary derived from botanical evidence. Isochrones for postglacial immigration of Spruce (*Picea abies*) into Fennoscandia, based mainly on ^{14}C dates. Modified after Moe, 1970, figure 1.

to justify dropping the Holocene on the grounds that the diachronism of the transition between glacial and postglacial conditions, unlike the onset of late-Cainozoic glaciation, is too blatant to be ignored (figure 10).[31]

Nevertheless some workers accept this deficiency with good grace and are content to envisage conflicting regional definitions. In one of the proposals that involves the demotion of 'Holocene' from epoch to age status, the Flandrian Stage of Europe, spanning the last 10 000 years, has the Aranuian as its counterpart in New

Zealand even though this encompasses 14 000–15 000 years.[32] The argument appears to be that it is premature to debate the issue of a common boundary; in contrast, C. B. Hunt simply dismisses the 'misconception that the Recent must begin simultaneously all over the world, although no other geological period did'.[33]

Palaeontological definitions

Hunt goes on to propose a return to the scheme proposed by Lyell in 1863 whereby the Recent would comprise deposits in which both shells and mammalian remains represent living species. Dissatisfaction with climatic criteria combined with a yearning for conformity underlie other moves in this direction, even if few would now agree with the view expressed by Keilhack in 1926 that the faunal difference between the Diluvium and the Alluvium is as marked as that between any other series.[34]

If Lyell's definition is reinstated the crucial faunal change will need spelling out. Large mammals are well represented in the 'Postglacial mass extinction' but their disappearance was not simultaneous and in some cases was delayed until late historical times; moreover no single species is available to serve as global indicator. And it is clear that, once various terrestrial and marine organisms are correlated for this purpose by reference to the environments they indicate, the boundary reverts to being climatic rather than purely palaeontological.

A plausible alternative is to make man the type species of the Holocene,[35] since 'both ecologically and culturally the end of the last glaciation marked a turning point in human evolution'.[36] Linnaeus had applied the name *Homo sapiens sapiens* to living Europeans;[37] were it to signify non-Neanderthaloid *H. sapiens*, the corresponding period would span 50 000 years and perhaps more. Since the requisite skeletal remains are scarce the range of datable deposits could be extended by reference to man's material remains, for example the end of the Magdalenian cultures in Europe. In justification one can always invoke Herbert Spencer's view of 'pyschosocial' development as one facet of evolution, and cite as stratigraphical precedent the use of fossil casts and animal trails by palaeontologists. Yet, as with other organic indicators, the correlation of regional equivalents (such as the Magdalenian) will in due course become inescapable; and correlating two such cultural divides with the help of the radiocarbon method amounts to little more than choosing an acceptable date. The alternative is to base correlation on a climatic or other environmental change. Unprofitable in itself (except as an extension of climatic methods) the operation goes full circle, so that the distinction between Palaeolithic and Mesolithic comes to be made 'by applying the quite arbitrary test to each industry: is it Late Pleistocene or Early Holocene?'[38]

A further variant of man-based definitions focuses on the biological and physical effects of his intervention, which could include the mammalian holocaust referred to above. Some Soviet workers claim that the Holocene is distinguished by the achievement of human society in Europe of such a high degree of development that it became a new and powerful geological factor; the resulting stratigraphical unconformity can be dated to between 10 000 and 15 000 years ago.[39] The difficulty lies in establishing when it was that man's influence was first felt. There are grounds for thinking that the use of fire by Palaeolithic man was no less influential than practices generally associated with agriculture. Again, the extent of man's contribution to mammalian extinction,[40] to soil erosion and the like, is still an open question. The various definitions given for the Anthropogene (a term used synonymously with Quaternary by many Soviet geologists) bring out the problems;[41] but a name which can be taken to mean both 'man-made' and 'giving rise to man' is perhaps destined to create confusion.

Arbitrary boundaries

The alternative to a wishful time-parallel boundary is one that is time-parallel by definition. We have seen that some proposals already amount to this: a chosen boundary is extrapolated beyond the type section with the help of quantitative dating methods. Since the values given by existing definitions range from 4000 to over 50 000 years ago, there is ample scope for choice; nevertheless a date in the region of 11 000 years ago would conform with many of the palaeoclimatic boundaries currently favoured.

Zeuner at one time proposed a round figure of 1 million years ago for the base of the Pleistocene, but subsequently decided that the advantages of such a boundary were outweighed by the practical difficulties it would cause in stratigraphical work. 'If . . . we cannot avoid adopting an artificial delimitation, it is best to choose one which necessitates the least number of changes in the conventional system of stratigraphy, and to define it by a feature which is not local.'[42] The feature cited in illustration by Zeuner is a radiation date, that is a date obtained by reference to a plot of fluctuations in the earth's intake of solar radiation and which therefore hinges on palaeoclimatic interpretation. This, together with a marked reluctance to disturb established stratigraphical convention, would severely detract from the one virtue of an arbitrary date, namely, its arbitrariness.

The suggestion that Precambrian times should be subdivided into equal units, such as megacenturies (each 10^8 years long), has been dismissed as of no value because the units have little or no connection with the rocks themselves.[43] Pleistocene–Holocene boundaries based on 'abstract divisions of non-material time' are viewed with disfavour for the same reason.[44]

Since geological time is not salami, slicing it up has no particular virtue. But if it is to be sliced there is no need to botch the job, and chronometric dating provides the guidelines. How compatible is the result with the use of named subdivisions of geological time? Wickman believes that year-based units of geological time eliminate 'the need for rigorous definitions of the boundaries between chronostratigraphic units. Instead, terms like Silurian and Mesozoic can be used in the same way in geology as concepts like Renaissance and Middle Ages are used in history. There is no need to find universal features that can serve as boundary lines'.[45]

This is a tempting solution for resolving the status and extent of the Holocene, with the added attraction that it would conform with the practice of using 'Pleistocene geology' to denote little more than a broad sphere of interest. Yet the historical examples given by Wickman are cautionary. 'During the Renaissance' is a restrictive phrase linked to the European phenomenon; few historians would relish speaking of Renaissance Patagonia. 'Middle Ages' has clearer temporal connotations; yet 'mediaeval Antarctica' is hardly a felicitous phrase. The implication is that, however loosely we define the Holocene, it will retain a 'postglacial' taint.

Names and 'time-bands' thus emerge as incompatible, at least as regards the present case. Let us now consider in detail the question of changes in sea level—a question which has acquired prominence in its own right—and see how it fares in the light of chronometric dating.

5 Ancient shorelines

> Where every object changes, it is difficult to find a measure of change, or a fixed point from which the computation may begin.
>
> John Playfair[1]

Changes in the relative elevation of land and sea have been accepted since men encountered sea-shells on mountains and recognised them for what they were. The concept of world-wide fluctuations in sea level was inherent in the biblical account of the Flood and hence underlay the many versions of Diluvialism, some of which held sway well into the nineteenth century. It was also manifest in the Universal Ocean of Abraham Werner (1747–1817), from which all the components of the earth's crust had supposedly been derived. But it was Edward Suess, in *Das Antlitz der Erde* (1883–1908), who first formally stated the eustatic theory, in essence the belief in synchronous swings of sea level over the globe.[2] Suess based his views on the geological evidence of widespread transgressions and regressions, and assumed that, with the exception of the mountain belts, the continents had remained stable throughout geological time.

Various mechanisms came to be proposed to account for the required changes in oceanic capacity. The earth movements they involved did not at first appear to undermine the postulate of overall continental stability. Even so, the Quaternary was especially attractive as a field for eustatic speculations because it was already endowed with an alternative source of sea-level change : the abstraction and release of waters locked up in ice sheets and glaciers.

Applications of a sea-level chronology

The potential clientele for a history of Quaternary sea level is a varied one. It includes the geodesist, for he is concerned not only with the present-day oceanic surface, by which he defines the geoid[3] and which serves as datum for his measurements, but also with fluctuations in its position, as these affect the earth's moment of inertia and are in turn affected by changes in the rate of rotation.

The needs of physical geographers and archaeologists are often more parochial. The interpretation of coastal features, whether physical (such as the products of erosion and deposition) or organic (peat deposits and anomalous coral banks, for example), requires knowledge of shifts in the position of restricted tracts of shoreline during the period in question. However, the submergence of sites and the disappearance of exploitable territory[4] are among the archaeological and historical problems that call for regional sea-level chronologies spanning recent millennia, although the use of sea levels as a means of dating ancient cultures throughout the world is not to be recommended.[5]

This brings us to the two fields in which sea-level history would seem destined to find its widest application : palaeoclimatology and stratigraphy. Certain coastal deposits such as coral or aeolianite, and the faunas associated with these and other littoral formations, may yield environmental information directly.[6] In addition, the sea levels they reflect are useful guides to the earth's glacial status. At present the volume of water held by the oceans represents 97 per cent of the world total compared with about 2 per cent in ice caps and glaciers, about 0.5 per cent in surface and subsurface continental waters, and 0.0001 per cent in the atmosphere.[7] Since 1842 —when Maclaren suggested that glaciation might have led to a fall in sea level of 350–700 feet (*c.* 100–200 m)—there have been many attempts to compute the eustatic effect of glacial fluctuations in the past and even in the future.[8] The converse calculation rarely goes beyond an assessment of the gross climatic conditions reflected by successive sea-level fluctuations; even so, the eustatic record often plays a prominent part in the appraisal of competing theories of climatic change.

The chief attraction of ancient sea levels for the stratigrapher lies in their world-wide incidence, whence the prominence given to the Flandrian transgression in the search for a Holocene boundary. Fossil beaches of known glacial affiliation can help in determining the stratigraphical position of related continental deposits both directly and in terms of base-level control. In the Mediterranean area, for example, the Last Interglacial beach has often served to demarcate fluvial deposits no older than the Würm glaciation,[9] while the altimetric correlation of river terraces with former sea levels was for long a regional speciality.[10] In addition, the climatic connotations of a particular sea-level position may be taken into account in the reconstruction of the corresponding terrestrial environments, so that (for example) the possibility of lower temperatures will be considered when interpreting sediments laid down during a pronounced marine regression.

The glacio-eustatic framework

The difficulties that attend the identification and measurement of fossil beaches are amply documented, as are those stemming from the distortion of the oceanic surface by winds, tides, atmospheric pressure, gravitational effects and so forth.[11] These and other technical problems can be minimised or at any rate borne in mind when reviewing the evidence.

The uncritical acceptance of glacio-eustatism is more severely limiting. Once high sea levels are automatically equated with interglacials and low ones with glacials, the freedom with which they can be interpreted is curtailed. This applies to the fossil as well as the inorganic evidence. For example, the marine faunas of some of the Mediterranean raised beaches apparently reflect conditions warmer than the present; 'cool' species found with them risk being dismissed as not being particularly diagnostic.[12] In the case of the Sicilian (90–100 m) beaches it has also been argued that the climatic requirements of the troublesome species may have changed; that the deposits in question were laid down during the early part of a glacial regression;[13] that if the sea was indeed cooled it was in response to physical events unrelated to glaciation, such as subsidence of the Wyville-Thomson Ridge or an irruption of the Atlantic over the Gibraltar sill;[14] and that the faunas reflect changes in salinity rather than in temperature.[15] The conclusion that 'The interglacial character of the high shore-lines of the Pleistocene,

therefore, can no longer be seriously challenged'[16] comes as something of a surprise.

Conversely it may be the elevation of a particular fossiliferous
beach which is regarded as anomalous. The epitome of a warm
association in the Mediterranean record is the Tyrrhenian fauna,
which includes the distinctive gastropod *Strombus bubonius*; the
fact that some *Strombus* beaches occur at or even below present
datum[17] shows that 'interglacial' sea levels can be found wanting
in height no less than in faunal character. Earth movements may
then be invoked to remove the anomaly. As we shall see, *ad
hoc* crustal distortion figures prominently in the glacio-eustatic
literature.

It goes without saying that homotaxial correlation is encouraged
if fossil shorelines acquire climatic labels. The matching of shoreline
'stairways' across the seas purely in terms of elevation has the saving
grace that it cannot disguise poor agreement. Once the factor of
elevation is eliminated by translating each succession into its
glacial/interglacial equivalent,[18] whether on the evidence of the
sea levels themselves or on their relationship to glacial deposits, the
result is a greater dependence on ordinal correlation.

Some sea levels have been given numerical ages which in turn
hinge on glacio-eustatic assumptions, as when successive interglacial
beaches are dated by reference to the Milankovitch curve of solar
radiation.[19] Radiometric methods have supplanted such procedures
and, what is more, opened the door to the testing of glacio-eustatic
models by the direct comparison of sea-level, glacial and climatic
chronologies. On occasion the result may conflict with expectations,
as when evidence for a low sea level yields a date normally allocated
to interglacial conditions.[20] All too often, however, such discrepancies are not allowed to emerge because high sea levels for which
dates are available tend to be accepted as interglacial in character
and the debate is limited to the age of these episodes.[21] Alterna-
tively, the issue may remain unresolved because not all the sequences
at issue are adequately dated. Thus, a eustatic curve recently constructed from a large number of sea-level indicators yielded seven
peaks which coincided with maxima on a palaeotemperature curve
derived from ocean cores; a rival time scale for the core record
places three of the peaks within glacial periods.[22]

Earth movements

By the early 1930s the straightforward application of glacio-
eustatism to geomorphology was being seriously challenged by a
growing awareness of continental instability. A decade ago it had
become apparent that Suess' model of eustatic movements superimposed on a largely rigid crust had 'seriously broken down, and
with it the possibility of large-scale cyclical correlation of landforms
on a height/time basis'.[23] The concept of isostasy played an important part in the reassessment. Attention was first focused on the
subsidence of land areas burdened by the ice sheets and on their
recovery during deglaciation; it was soon realised that comparable
depression and rebound were likely to characterise the sea floor
following the release and abstraction of water by the ice bodies.
Hence changes both in the elevation of the land and in the capacity
of the oceans would stand in the way of a simple conversion of a
glacial fluctuation into its eustatic equivalent. With the growing
acceptance of sea-floor spreading the situation has become even
less favourable to a purely climatic interpretation of sea-level
history, as the resulting modifications of oceanic volume (unlike
those of isostatic character) cannot be correlated with glacial
episodes.

Nevertheless the search for a generalised, global eustatic curve
continues. The eustatic record may even be employed to bring out
crustal movements : for instance, the fact that successive interglacial
sea levels are progressively lower in elevation is commonly taken to
indicate that glacially-controlled oscillations have been superimposed on a secular increase in the capacity of the ocean basins.[24]
But for this approach to be feasible, stable areas must first be
identified.

There is no accepted definition of stability. Like contemporaneity,
though perhaps with more justification, the term is used in a relative
sense. The Canadian and Scandinavian Shields rank as stable in
certain seismological contexts; to the student of isostasy they are
distinctly mobile. It is not, however, simply a question of the length
of time under consideration : stability is often found to have been
illusory, and no portion of the crust can be assumed to have remained entirely immobile throughout the Quaternary.[25] Even so,

sea-level chronologies based on relatively stable areas are still used as the standard by which deformation elsewhere can be measured. In 1802, prior to the introduction of ice sheets and glaciers into the eustatic equation, John Playfair commented :

> The imagination naturally feels less difficulty in conceiving, that an unstable fluid like the sea, which changes its level twice every day, has undergone a permanent depression in its surface, than that the land, the *terra firma* itself, has admitted of an equal elevation. In all this, however, we are guided much more by fancy than reason; for, in order to depress or elevate the absolute level of the sea, by a given quantity, in any one place, we must depress or elevate it by the same quantity over the whole surface of the earth; whereas no such necessity exists with respect to the elevation or depression of the land.[26]

Glacio-eustatism has not removed the temptation to mobilise the *terra firma* when all else fails.

The alternative is to obtain the standard from areas whose instability is well documented.[27] It can be argued that to know the measure of one's foe is half the battle. What is more, both uplift and subsidence favour the survival of eustatic evidence. The American Gulf Coast and the Low Countries can justify their prominence in the sea-level literature on this count, as their subsidence has promoted the formation and preservation of blanket peats, estuarine deposits and other sediments rich in datable material; conversely the progressive rise of Scandinavia and of Barbados has lifted successive shorelines safely above the waves. Nevertheless we are still left with the problem of evaluating the movements, and it is difficult to resist turning to the beaches themselves for some of the information.

The Flandrian transgression

Some of the studies devoted to the eustatic history of recent millennia are explicitly concerned with comparing the observed sea-level changes with those predicted by glacial chronology and associated hydrological and geophysical events.[28] Many are seemingly motivated by an interest in sea levels *per se*. This does not reduce their value for other fields of research, the real need being for continual reassessment of the 'accepted' chronology. In this respect it is noteworthy that studies of changes in the earth's rotation give prominence to evidence which many geologists would treat with considerable reservation for a postglacial sea level higher than the present.[29] The same evidence has been invoked to support the 'contemporaneity' of a Climatic Optimum over the globe.[30]

It is generally accepted that a major regression accompanied the last glacial stage and that deglaciation brought sea level to its present position. Several recent estimates put the magnitude of the regression at 120–130 m. The ensuing Flandrian (or Holocene) transgression was apparently marked by halts and perhaps by temporary reversals.[31] Where opinions conflict most sharply is over the closing stages of the eustatic rise.

Three main schools of thought have been recognised.[32] The first maintains that the sea rose to its current level 3000–5000 years ago and has since fluctuated above and below it, the second that sea level has remained relatively stable since that time, and the third that the rise has been continuous. In some respects the viewpoints diverge less than suggested by a drastic classification of this kind. For example, many exponents of the third envisage a deceleration in the rate of rise between 6000 and 3500 years ago, which provides opportunities for a concordat between them and the 'steady' school; indeed it is sometimes impossible to decide between the two theses when evaluating the field evidence.[33] Furthermore, most published curves agree in showing that the sea lay between 8 and 12 m below its present level about 7000 years ago.[34] However, the divergences remain striking even within restricted areas (figure 11).

The changes that have taken place since historical times are also uncertain, not even the period of instrumental record being immune from doubt. Several authors detect a general rise over the last hundred years or so but disagree over its magnitude and rate, and there is some discordance between the evidence in different parts of the globe for the first decades of the present century.[35] This has not cramped attempts to correlate the record with phases of volcanicity, sunspot cycles, tree-ring sequences and other chronologies which are not in themselves entirely secure; furthermore the validity of glacio-eustasy as a whole has been staked on the evidence of instrumental observations.[36] It may perhaps be wiser to regard the search for a global average as 'a marginal undertaking',[37] especially

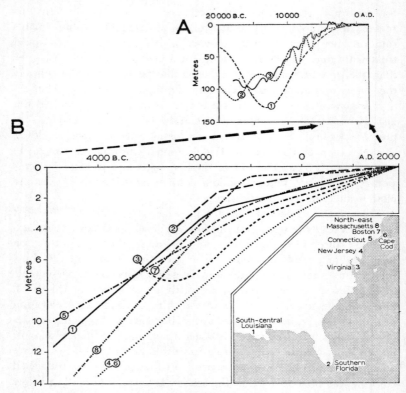

Fig. 11   Sea-level curves. (A), Global, since 20 000 B.C. (1) after Milliman and Emery, 1968, figure 1; (2) after Curray, 1961, figure 1; (3) after Fairbridge, 1963, figure 5. (B), Eastern USA, since 5000 B.C. Various authors, modified after Scholl and Stuiver, 1967, figures 7 and 8.

since the longest tidal-gauge record available—that of Amsterdam, which goes back to 1682—applies to an area renowned for its subsidence.

For the record of earlier periods reliance initially had to be placed on conventional stratigraphical methods, whereby the sea-level indicator was related to marine and continental deposits of known geological age, and on archaeological content and context. Relative freshness was sometimes found a useful guide to the antiquity of adjacent features.[38] Ancient harbours, early maps and other historical indicators served for the recent past.

All such guides to the changing elevation of the land/sea contact call for a measure of interpretation. Radiocarbon dating has merely shifted the source of uncertainty. If the method is applied to organic remains, their relationship to the contemporaneous sea level has to be established even when they are manifestly in place. The use of shell material has been criticised because of the risk that it was carried to its present position by storm waves, birds or natives.[39] The dated material or that underlying it may also be distorted by oxidation, drying out and the like; in Connecticut and the Netherlands, Holocene deposits have been reduced to as little as 10 per cent of their original thickness as a result of compaction.[40] The risks of sample contamination are high.[41]

By the same token the role of earth movements is not always easy to isolate. In south-central Louisiana a series of peats ranging in age between 7240 and 1475 years was found to show that sea level had attained its present position 3650 years ago, it being assumed that subsidence had prevailed at 7 cm a century. Once the correction factor for subsidence is halved the same evidence indicates that the sea has continued to rise.[42] The construction of eustatic curves from data drawn from many parts of the world might appear one solution to the problem, in that local deviations would tend to cancel one another; but there is a compensating risk that the coincidence of uplift or depression in several areas will give rise to spurious fluctuations in the sea-level trace. In one widely reproduced curve, low sea-level stages refer to evidence from low-lying areas which include deltas prone to subsidence, and the high stages to data from coasts close to important mountain ranges which might be suspected of undergoing uplift.[43] Appeals to local deformation enable anomalous points to be harmonised with the expected curve;[44] conversely, conformity may lead one to disregard the possibility of distortion in areas (such as Japan[45]) where recent instability is manifest.

As we have seen, it is hazardous to label any coastline as inherently stable or unstable. The absence of raised beaches higher than 3 m in Madagascar—considered a stable area—has as corollary large-scale Pleistocene warping of thousands of miles of coast in territories of lower repute where such beaches are to be found.[46] Conversely, traces of recent emergence by a few metres have led to the postulate of little or no uplift or subsidence during the last 3000–4000 years in Africa, southern Europe, the east coast of the Americas from New Jersey to Argentina, India, Australia, New Zealand, and oceanic islands by the hundred, whereas the lack of such traces in Nova Scotia, Newfoundland and British Columbia demands local sinking.[47]

In consequence, the evidence may be in harmony with more than one view of recent sea-level history. At Tealham Moor, Somerset, the base of a peat deposit has been dated to 5412 years ago. It is the sole item at sea level shown on one of the graphs supporting the steady view.[48] The alluvium on which the peat rests was allegedly laid down below and up to mean sea level; another interpretation is that it formed between mean sea level and high tide and that the peat is consequently a poor indicator.[49] We meet Tealham Moor again on a 'postglacial high sea level' curve, to which it is accommodated by invoking a temporary marine regression.[50] In this role it is accompanied by a date from Takoradi Harbour, Ghana, which ranks here as 'zero sea level' and in another publication as a '2–4 ft raised beach'.[51]

Although it was to be expected that curves based on data from restricted areas would disagree with one another, the discrepancies between all the published global schemes have aroused some impatience. In 1967 an expedition was mounted with the aim of resolving the issue so far as it concerned some of the Pacific atolls. These had supplied much of the support for the hypothesis of a Flandrian sea level higher than the present, the evidence being largely morphological with a few supporting radiocarbon dates. The expedition found no proof of submergence during the last 7000 years in the Caroline and Marshall islands. The data collected in the eastern Carolines indicated a continuing rise of sea level which slowed down markedly about 4000 years ago.[52] As regards the southern Marshalls it could only be concluded that sea level 2500–3000 years ago had lain close to its present position. Guam displayed elevated reefs but these could be attributed to recent uplift.[53]

This work, with its emphasis on the careful collection of samples for radiocarbon and ionium–thorium dating, demonstrates the value of a concentrated attack on a relatively restricted area in order to test a particular hypothesis. Yet its results are not entirely satisfactory. No general conclusion could be reached as to the closing stages of the Flandrian rise—that is to say whether the steady view was preferable to the 'continuing though decelerating' one. The raised terraces of Guam were dismissed on the grounds that the island lies north-west of the Mariana Trench (which is characterised today by a good deal of seismic activity) and that it displays several high terrace levels which are clearly not due to sea-level changes. In brief, the scope for special pleading remains greater than one might have hoped.

The eustatic–tectonic dichotomy

It is hardly surprising that murmurs of revolt should be heard. The idea of coastal stability has been described in some quarters as a myth which casts doubt on all attempts to define a truly eustatic curve of sea-level rise,[54] while a eustatic curve of global applicability comes to be dismissed as chimerical.[55]

If the land cannot serve as reference datum for sea-level changes, we are left with two options : the centre of the earth, or some hypothetical surface of known shape and dimensions surrounding the globe such as the spheroid against which the form of the geoid is usually matched. Neither is as yet a practical proposition. The only conclusion would seem the need to plough the old furrow and hope that more and better dates and stratigraphical controls will progressively circumscribe the area of conjecture.

But the question goes deeper. In 1901 Pearson termed the choice between movements of the land and those of the sea 'the greatest problem of geology, or of geography', and drew attention to the signal lack of progress in settling the matter in the preceding two thousand years.[56] The last seven decades have revealed not only that the earth's crust is extremely mobile but also that it is responsive to what had once seemed insignificant external forces. Gill has argued that 'It is a false antithesis to ask whether a given drowned or emerged coastline is a function of tectonics or eustasy. Both factors are involved and it is generally a matter of determining whether a particular factor is of significant magnitude'.[57] Could it not be that the factors are destined to remain inseparable, and that, as with the mechanist–vitalist controversy or the 'nature or nurture' debate,[58] we are in the realm of badly posed questions and false dichotomies?

The value of an authoritative, up-to-date, generalised eustatic curve has already been stressed. It is needed both by nongeologists and by geologists investigating areas for which the record remains unknown. The didactic benefits of isolating the eustatic factor are clear. At the same time there is room for local studies which exploit the local record—warts and all—to advantage. One possibility is to use well-dated sea levels as reference planes against which terrestrial deformation can be measured. Not least among the virtues of this approach is that it diverts the abundant data on Quaternary shorelines away from the narrow issue of glacio-eustatism to the wider field of crustal deformation and in particular to the testing of models of the earth's interior.

Shorelines as reference surfaces

Ancient beaches have long been employed as a guide to crustal warping. In G. K. Gilbert's classic study of Lake Bonneville, published in 1890, successive lake terraces, though valuable as indicators of changing climatic conditions, found their most fruitful application in terms of how far they deviated from horizontality. This measure, once allied to estimates of the corresponding lake volumes, enabled the budding concept of isostatic compensation to be tested. At its greatest extent Lake Bonneville was almost as large as present-day Lake Michigan; when it shrank to form the salt lakes that now occupy part of the basin, it left a number of shorelines. Gilbert found that these were upwarped; his crucial observation was that the greatest uplift had occurred where the lake had been deepest. Recent work has corrected Gilbert's observations, but his geophysical interpretation remains valid.[59]

The sequence of lake levels is now known to have been much more complex than Gilbert believed. An important event was the diversion of the Bear River into the basin by a lava flow, a finding which strengthens the case for a nonclimatic interpretation of the evidence and for the establishment of a chronology that does not hinge on climatic correlation. The rate at which the lake shrank is clearly central to any interpretation of the crustal response to unloading. Fortunately radiometric dates are available for the latter part of the sequence.[60]

Many studies of recent crustal warping, notably in Scandinavia and North America, exploit sea-level evidence for similar ends. In some of them an attempt is made to compute total uplift or depression by allowing for global eustatic changes; in view of the problems this raises it would seem safer to concentrate on the relative displacement of individual beaches relative to modern sea level or to one another. Even this calls for assumptions about the extent to which the former ocean surface was deformed by tidal, gravitational and kindred factors, but there is no reason why suitable corrections should not be attempted.

If attention is restricted to small stretches of coastline it may prove possible to identify and trace individual beaches directly by reference to landforms and deposits, but the progressive character of transgressions and regressions means that the feature in question need not necessarily represent an isochronous time-plane and that its apparent deformation or seaward tilting could be nothing more than an expression of the mode of formation. Where the evidence for the postulated shoreline is discontinuous the need for multiple dating is even more obvious : a single determination on a fossil beach merely establishes the age of the point that has been sampled.

Of the many Quaternary beaches that have served to reconstruct the isostatic recovery of Scandinavia, the *Tapes* shorelines come closest to the ideals expressed above in that their recognition depends in part on the presence of andesitic pumice at numerous exposures. Unfortunately the feature represents a complex of shorelines rather than a single, brief episode. Furthermore the pumice (which is also present in beaches lower than the *Tapes* shorelines) is known to include material which cannot be attributed, as was formerly thought, to a specified eruption in Iceland about 4000 years ago.[61] Nevertheless pumice is beginning to play a useful role in strandline investigations in Canada,[62] and in Spitsbergen it has enabled the relative slope of two successive water-planes to be determined (figure 12A).[63] Granted the limitations imposed by lack of information on the time elapsed between eruption and the deposition of pumice on the shore, the method is clearly of great promise.

The dating of beach features by archaeological and palaeontological remains has also found wide application in Scandinavia. The usual reservations apply. Indeed, faunal changes are more than usually prone to diachronism in areas undergoing deglaciation, as both terrestrial temperature and the salinity and temperature of

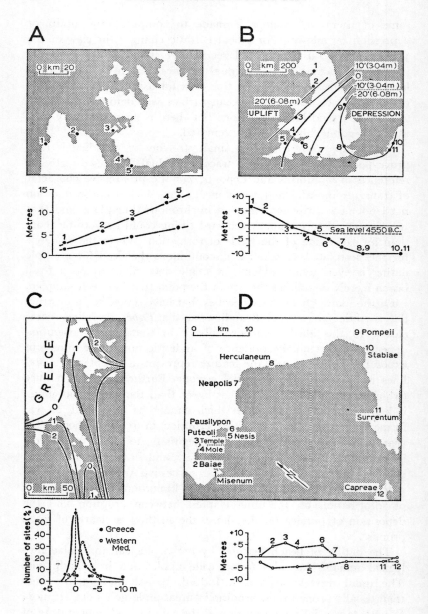

the surrounding seas will exhibit steep gradients. The conflict between the shoreline representing the highest limit of the *Littorina* Sea of the Baltic and that marked by the contact between salt-water and freshwater *Littorina* deposits is understandable.[64] Again, there is some link between the distribution of archaeological sites and the postulated shorelines, but since many of the sites are dated by reference to the beaches the system lacks stability.[65] Even where the sites are dated independently the fact remains that their relationship to sea level during occupation will remain conjectural unless they include harbour or analogous structures. Pollen analysis and varve-counting have proved valuable in preliminary investigations, but the uncertainty introduced by the correlations they demand renders them inferior to radiometric methods.

Few shoreline studies employ carbon-14 dating on a scale that accords with the requirements of a chronometric approach. One of them is the work of J. T. Andrews on postglacial uplift in Arctic Canada.[66] It departs from the procedure advocated here in employing strandline features dated by carbon-14 principally to check the predictions obtained from uplift curves, although these are based on the age of the marine limit at various localities. On the other hand Andrews also uses emergence curves to reconstruct episodes of synchronous shoreline activity. The operation embodies the principle that 'water-planes' can be reconstructed in the absence of recognisable strandline features provided there are sufficient dated sea-level indicators at a number of coastal sections, the hypothetical shoreline being delineated by linking points similar in age. The shorelines provide a measure of the displacement undergone by

Fig. 12   Dated sea levels as guides to crustal deformation. (A), Shoreline relation diagram of two raised beaches in North East Land, Spitsbergen, dated by drifted pumice. Modified after West, 1968, figure 8.6b. (B), Displaced sea-level indicators dated to 6500 y.a. in Britain and Netherlands. Modified after Churchill, 1965, figure 1. (C), *Upper*, contours of best-fit third-order surface for rate of downward displacement in metres per 1000 years of 23 coastal archaeological sites in the eastern Peloponnese during the last 2000 years. High rate of variation in south-west coincides with a heavily faulted zone. *Lower*, percentage frequency of occurrence of sites plotted against displacement relative to present sea level over the last 2000 years. Modified after Flemming, 1968, figures 1 and 2. (D), *Upper*, Roman sites in the Bay of Naples; *lower*, elevation of highest and lowest marine erosion marks on Roman buildings, indicating maximum amplitude of earth movement near *Baiae* and Pozzuoli (*Puteoli*). Modified after Flemming, 1969, plate 2.

different parts of the area of study. The method is also illustrated by the work of D. M. Churchill on the North Sea.[67] Churchill identified eleven peat deposits which had apparently formed at sea level 6500 years ago, nine of them in Britain and two of them in the Netherlands. Their elevation relative to present sea level was thus a measure of cumulative uplift or depression since their development (figure 12B). It so happens that their displacement supports other evidence for a progressive south-eastward tilting of Britain coupled with subsidence of the North Sea Basin;[68] in the present context what is significant is that, once the relationship of the samples to the contemporaneous sea level had been established, their littoral character lost its importance.

The work by N. C. Flemming on historical changes in sea level embrace a far larger area, namely the western Mediterranean Basin.[69] Their primary aim was to determine eustatic changes during the last 2000 years. By making detailed surveys of Greek, Roman and Phoenician coastal sites Flemming was able to establish their original relation to sea level and hence the extent to which they had been 'displaced' since their occupation. In other words the form of the sea-level surface for the period of 2000 years ago was compared with that of its modern counterpart. Flemming concluded that, within a working accuracy of ± 0.5 m, there has been no net eustatic change. Nevertheless some sites are displaced, and Flemming observes that they lie either on deltas or in volcanic or seismic zones, the Naples area being a case in point (figure 12D). Detailed analysis of the Aagean evidence shows that the postulated deformation is in good agreement with what is known of the area's tectonic history (figure 12C).[70] It is to be hoped that a similar study will become feasible for successive phases in the history of the Mediterranean during prehistoric times, so that long-term trends in crustal deformation can be established.

6 Cores and boreholes

A dark
Illimitable ocean without bound,
Without dimension, where length, breadth and highth,
And time and place are lost.

Paradise Lost

Dated sea levels lend themselves to the study of crustal deformation because they provide a horizontal datum at a specified time in the past. Time-planes derived from other sources are often 'deformed' from the outset, or at any rate are not diagnostic of horizontality. Consider, for example, the outcome of plotting the present-day land surface, that is to say the time-plane for 0 years ago.

By concentrating on the evidence yielded by boreholes and cores, this chapter emphasises the problems attendant on time-correlation initially bereft of altimetric connotations. The depths exposed in open sections are insignificant when compared with those attainable by drilling, but sections have the advantage when it comes to tracing individual erosional or depositional features laterally, or when the aim is to determine the variability exhibited by such units. As high costs generally rule out a dense network of boreholes, subsurface studies tend to involve an even greater reliance on 'blind' correlation than prevails in the study of exposed deposits. Of course boring may serve as an adjunct to direct observation, for instance when employed to reconstruct the shape of a valley now partially buried by alluvium or the extent of a fossil soil first recorded in a river bank. But there are times when inaccessibility rules out any check on correlation beyond that provided by the indirect findings of a geophysical survey.

Two fields which draw heavily on core data will serve to illustrate the discussion : palynology and the study of deep-sea deposits.

Treating them jointly is further justified by the prominent part they play in the investigation of Quaternary climatic change.

Pollen analysis

The difficulties that hamper the reconstruction of vegetational history from pollen sequences include variations in the amount of pollen produced by different species, in the efficacy with which it is dispersed, and in the extent to which the different components are preserved at the site cored. Additional problems are created by the widespread practice of presenting the pollen spectrum for each horizon sampled in terms of percentages, as variations in one species or plant group may thereby give rise to spurious changes in the representation of other components.

These problems are well understood and attempts have been made to counter them.[1] For example, the interdependence of percentage values can be eliminated by expressing each component of the pollen spectrum in terms of its net accumulation rate. For this to be possible it is essential to determine the rate of sedimentation throughout the section under review by obtaining a closely-spaced sequence of radiocarbon dates; pollen numbers can then be given as successive totals per unit area (figure 13).[2] In contrast, little can be done to establish the source area for the sample as a whole. The issue is a familiar one to those intent on discovering the provenance of particulate atmospheric pollutants, although they have the advantage of dealing with known points of emission and with documented atmospheric conditions.

It is thus understandable that palynology should be more often concerned with the regional than with the local situation. Some workers have profitably used pollen to investigate the environment of ancient sites and the effect of man on the vegetation cover.[3] The regional trend may then serve as a standard by which to gauge the extent of human influence.[4] But it is often found preferable to use independent sources in assessing man's role in vegetational change in order to allow for it in the search for the 'natural' situation. Other local influences on plant growth, such as fluctuations of the water table, can also be taken into account by reference to the deposits from which the pollen was recovered and other environmental indicators.

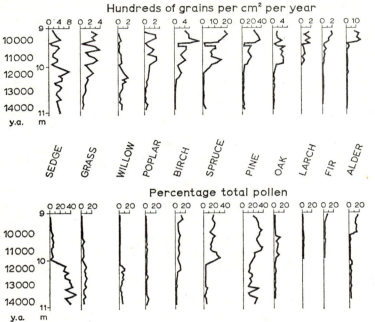

Fig. 13 Pollen deposition in Rogers Lake, Connecticut, between 14 000 and 10 000 years ago. Upper curves show accumulation rates of major pollen types plotted against depth (m) and age. The lower curves show conventional plots expressed as percentages of total terrestrial plant pollen. Modified after Davis and Deevey, 1964, figure 3.

By good fortune a regional approach accords well with the interest in climatic reconstruction that has long characterised pollen studies. A strong impetus in this direction was doubtless given by the rise of modern palynology in Sweden in the second decade of this century, the chronology of deglaciation being uppermost in the minds of many of those responsible for perfecting techniques of collection and analysis. The corollary is climatic correlation, first between cores and then between regions, and this in turn reinforces the need for converting the evidence to units which bear some relation to the climatologist's zones.

Zoning is also applied 'vertically' to the pollen sequences. The resulting subdivisions amount to biostratigraphical units, a familiar example being the nine pollen zones which are commonly recognised in the last 14 000 years of the north European record. The boundaries are not always self-evident, because even on percentage diagrams the various components of the spectrum do not always display pronounced fluctuations unanimously. The palynologist must therefore seek out features that denote noteworthy climatic changes. Correlation between such zones can then be made on the basis of analogous or complementary trends, it being clear that the effects of a particular climatic episode will differ from place to place. A possible drawback is that, should an episode which is widely represented be missing from a new core, there will be a reluctance to accept this absence simply as an indication that the climatic episode was not locally felt, and the gap may come to be ascribed to an insensitive vegetation.

The success of using pollen percentages for stratigraphical ends has been ascribed by one author to the homotaxial character of the zones thus defined,[5] but it is more usual for the time-transgressive nature of pollen zones to be dismissed as not being serious. As a result, radiocarbon dates obtained from a particular zone in one core are sometimes extrapolated with the help of pollen analysis far afield. In some instances additional dating shows that the extrapolation was justified, in others that it was not.[6] The usual reservations regarding time-correlation on biostratigraphical and climatic criteria clearly apply.

It turns out that the extreme ease with which pollen can be recovered, identified, and counted is one of nature's more seductive tricks. It has permitted a valid and useful stratigraphic division of late-Pleistocene deposits, now largely superseded by radiocarbon dating, while the changing pattern of vegetation remains as elusive as ever.[7]

Marine cores

How far does this apply, *mutatis mutandis*, to marine sediments? The question is a reasonable one because, despite the wide range of motives behind the collection and analysis of cores taken from the sea floor,[8] their chief attraction for the Quaternary geologist lies in the promise they hold of a well-preserved record of temperature fluctuations.

As in the case of pollen, areal and temporal zoning of the evidence is both a consequence and a prerequisite of palaeoclimatic interpretation. Comparisons between the present-day and the former distribution of planktonic Foraminifera, for example, call for the delineation of boundaries between species or assemblages thought to differ in their environmental requirements. Again, the reworking of bottom sediments by burrowing organisms means that only the major temperature fluctuations will be detectable, a limitation not displeasing to the seeker after climatic episodes of global significance. It is noteworthy that here, too, both kinds of zone tend to be drawn on the basis of percentage counts, with the result that the representation of diagnostic species may be distorted by more tolerant ones.

What the recorded species are diagnostic of is not always certain. The area of doubt is increased when the modern distributions that serve as the basis of comparison are obtained from the topmost parts of the cores as opposed to the ocean surface; and doubt yields to perplexity when one discovers that some of the planktonic species whose shift serves to reconstruct mean differences between modern and ancient temperatures of less than 5°C move daily through a range of 10–12°C, while most of them have distributions related to water masses which are identifiable in terms of temperature–salinity (TS) relationships and whose links with the local climate are not immediately apparent. Changes in the coiling direction of individual species, sometimes used as an alternative to more conventional biostratigraphical criteria, are also poorly understood. We may therefore sympathise with the conclusion that attempts to interpret 'colder water' faunas as representative of glacial or subglacial stages are often speculative and 'probably should be regarded as interesting possibilities'.[9]

The biological evidence may nevertheless be strengthened by the results of physical and chemical analysis of the sediments. For instance, under certain conditions calcium carbonate content is apparently temperature-dependent; the presence of ice-rafted debris is another clue to surface conditions at the time of deposition. In this context the greatest claims are made for the oxygen-isotope method, which depends on temperature-controlled changes in the oxygen-16/oxygen-18 ratio in calcareous tests of planktonic Foraminifera. Unfortunately there is dispute about the degree to which this ratio is a reflection of the temperature of the surface waters, and some workers believe that changes in it are largely controlled by the addition and abstraction of water from the oceans by ice sheets and glaciers. The process would also affect salinity, and the resulting faunal movements could well counterfeit the shift of climatic belts.[10]

Whether or not these and kindred difficulties are acknowledged, the core zones that emerge will invite correlation in order to eliminate the gaps and zones of mixing that mar individual cores and hence to obtain sequences applicable to zones hundreds if not thousands of kilometres wide. The hope is that, ultimately, long cores tested by cross-correlation will make it possible to establish a standard Pleistocene section. For, if 'the climatic changes of the Pleistocene were of world-wide effect, the corresponding faunal zones in deep-sea sediments ought to be correlatable over long distances', and, as it happens, correlation among many cores from widely scattered stations in the Atlantic and Caribbean establishes the validity of several such faunal zones.[11]

Correlation will also serve to link marine sequences with those obtained on land. Indeed, deep-sea cores are expected to provide a chronological framework into which the fragmentary record of the land can be fitted[12] and which will enable the age of terrestrial events beyond the reach of radiocarbon dating to be determined.[13] In similar vein, the value of the isotopic record of the oceans is held to be enhanced by the distortions stemming from the waxing and waning of ice sheets, on the grounds that it provides us with a time-sequence of glacial events on land.[14]

There is no obvious reason why the marine record should not be subordinated to the needs of continental stratigraphy. Yet the adoption of the binary glacial/interglacial code has two drawbacks. First, it reduces the yield of information inherent in the data. Second, it promotes the correlation of cores with one another and with continental sequences on an ordinal basis, on the assumption that fluctuations represented in one area will find counterparts elsewhere. The scope for personal judgement is thus considerable. To take an example, where one school identifies the Mindel–Riss interglacial of northern Europe with a foraminiferal zone spanning about 600 000 years, a rival correlation makes it correspond with an isotopic stage which apparently lasted about 50 000 years (figure 14).[15]

It will be interesting to see how the study of cores taken from

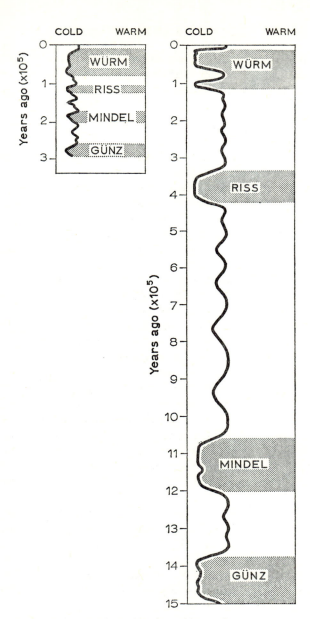

Fig. 14 Correlation of palaeoclimatic evidence from ocean cores with glacial stratigraphy. *Left,* curve based on oxygen-isotope palaeotemperatures from low-latitude cores. Modified after Emiliani, 1955, figure 15. *Right,* curve based on planktonic Foraminifera from Atlantic cores. Modified after Ericson and Wollin, 1966, figure 23.

ice caps develops. All the indications are that it will conform to stratigraphical convention and that pride of place will be given to the investigation of climatic history.[16]

Time-correlation between cores

Many cores yield datable material only in their upper portions. The usual sources of error are sometimes compounded by those stemming from a lacustrine or marine environment,[17] but the application of alternative radiometric methods and the dating of different components of the deposit will often reduce the problem to manageable proportions. An interim device which has found wide favour is to accept the apparent age of the uppermost sample as an empirical value for the excess ages to be expected throughout the core; thus, in the pollen study of Rogers Lake mentioned earlier, all the radiocarbon ages were reduced by 770 years as this was the measured age of surface sediment on the lake floor. The loss of resolution produced by burrowing organisms is not easily countered, although one can at least guard against it by paying due attention to the sediments in the crucial sections of the core.

Such precautions are of further value in drawing attention to the provisional nature of the dates. It is easier to overlook the assumption that governs the extrapolation of these dates to the lower parts of the core, namely that the rate at which its sediments accumulated has remained constant. To be sure, cross-checks using the geopolarity scale show that average sedimentation rates at individual deep-sea cores have remained fairly constant over long periods.[18] Even so, the legitimacy of taking this for granted is dubious.

Questions of rectitude apart, there is much to be gained both in oceanography and in limnology from a knowledge of regional and temporal variations in the rhythm of sedimentation. For example, studies of crustal downwarping where sedimentary loading is thought to have contributed to the movement require data on the rate at which the load was applied; actualistic models of geosynclinal development draw on measured sedimentation rates to illustrate the proposed mechanism; the student of soil erosion may wish to ascertain whether the history of lake deposition corresponds with the changes in sediment yield he has determined for its catchment.

Core zones have already been pressed into this service, the procedure being to divide their thickness by their estimated time-span. More insidiously, they may serve as a vehicle for reducing a series of determinations to a smaller number of average rates for the various zones. It is on this basis that sedimentation rates higher in the Pleistocene than in the Holocene are reported from many parts of the ocean. The finding accords well with the rapid mechanical weathering, glacial erosion, lowered sea levels and other characteristics of glacial times likely to promote high sediment yield to the sea; but if we are to go beyond this level of generalisation the links between changes in sedimentation and the possible controls (whether continental or oceanic) will need to be made more explicit.

It follows that a single core, however well documented, is doomed to inadequacy. Take the finding that in a part of the mid-equatorial Atlantic the rate of clay deposition changed from 0.22 to 0.82 $g/cm^2/year$ about 11 000 years ago, and that there was a corresponding drop in total carbonate deposition from 2.80 to 1.34 $g/cm^2/year$: as its authors conclude, the sympathetic change cannot be interpreted with conviction without extensive studies of the circulation patterns and sediments of the Atlantic as a whole.[19] In contrast, a closely-spaced network of cores armed with dates can be linked by means of time-planes (figure 15) whose divergence or convergence can then be related to possible controls of deposition. Allowance will have to be made for reworking by bottom organisms, and for core-shortening especially if a gravity corer is used; again, the time-planes are likely to need revision once fresh cores and additional (or better) dates are obtained. Even so, interpolation (as opposed to extrapolation) provides some safeguard against gross error.

The time-planes are unlikely to bear any simple relationship to stratigraphical boundaries. This is not to deny that dating such boundaries can be a productive exercise provided evidence of diachronism is accepted as a potential source of fresh information rather than as a violation of cherished tenets. Let us reconsider in this light our review of attempts to define the Pleistocene–Holocene boundary. Reference to pollen zones yields an age of about 7000 B.C. in parts of Sweden and one of 8300 B.C. in Denmark.[20] The analysis of ocean cores gives a date of 6825 B.C. for the northern Indian Ocean and one of 14 550 B.C. for the

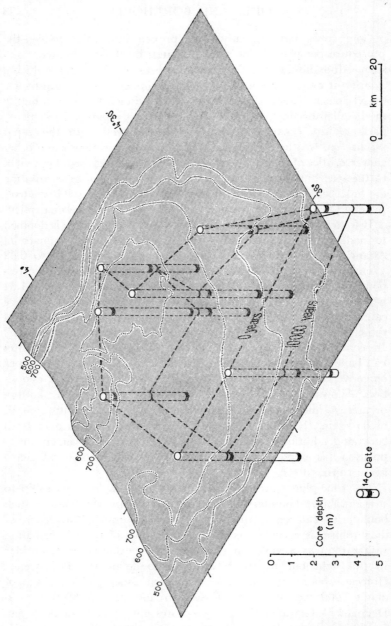

Fig. 15 Radiocarbon isochrones as indicators of areal variation in rates of sedimentation on the sea floor: an example from the western Mediterranean. Modified after Huang *et al.*, 1972. Depths are in fathoms.

Caribbean,[21] while no obvious boundary can be detected in some Arctic and Antarctic cores spanning the last 70 000 years.[22] Where the aim is to obtain an acceptable global date, the disparities and gaps are a nuisance; yet they are a boon to anyone seeking to trace the spread of plants (*see above*, p. 44) and other organisms as the ice retreated. One can go further. The close of the Atlantic period was originally defined in Britain by a marked decline in Elm pollen. Radiocarbon dates were later obtained for an analogous feature on pollen diagrams for many parts of north-western Europe; the dates lie within one or two centuries of 3000 B.C., a scatter which is regarded as too great for the decline to be a climatic effect, and the intervention of Neolithic farmers (already implicated by the presence of the pollen of plants associated with man) is seen as a more satisfactory explanation.[23]

Nevertheless a preoccupation with boundaries may lead one to overlook authentic cases of contemporaneity. Whereas the Zone III/Zone IV contact obtained from pollen cores in northern Pacific America appears to be some 1500 years younger than its equivalent in southern Chile, palaeotemperature curves derived from the same vegetational records show considerable agreement between the two areas.[24] And it is as well to bear in mind that local conditions relating to soil, drainage and the like can give rise both to 'synchronous plant successions which are difficult to detect as parallel, and to apparently similar successions which are not necessarily closely synchronous'.[25]

The wider view

At some stage in the investigation the chronometric framework will have to be extended to encompass the source areas for the material making up the core sample. There can be no question of seeking 'key' palaeoclimatic tie points to link source and destination even if this is the course prompted by a lack of radiometric control.[26] An independently dated fluvial chronology is more helpful to the interpretation of offshore terrigenous sedimentation than one founded wholly on hypothesis; in the Mediterranean Basin the alluvial record in fact points to a succession of periods of high and low sediment yield which bears little relation to what glacial chronology might lead one to expect.[27] Similarly, surveys of present-day pollen deposition coupled with palynological analysis of dated fluvial sediments help to define the extent to which surface runoff has collaborated with wind in transporting the pollen to the area of study.[28] There may be incidental gains. For example, isotopic analysis of marine Mollusca from a series of dated deposits in the coastal cave of Haua Fteah, in Libya, yielded a much higher resolution than would have been possible with samples taken from a few centimetres of marine sediment representing the same period.[29]

On the other hand, ambiguous evidence may be prematurely resolved in the search for a harmonious general picture. Ash erupted by Santorini about 3370 years ago has been found to have a refractive index similar to that of the 'upper tephra' recorded in cores taken in the eastern Mediterranean, and mapping the ash fall apparently gives a picture consistent with the suggestion that the eruption had been catastrophic for the Cretan population even though there is little trace on Crete of either the ash or the violent waves that should have contributed to the destruction. Doubts have already been expertly cast on the chronology of the tale as a whole (not to mention its identification with the Atlantis myth);[30] our concern is with the dates for the tephra. Core V10–58 yielded a radiocarbon date of 8600 years for a deposit under the tephra and one of 5300 for material above it. In core V10–52 the ash is overlain by a layer which has been shown in a third core to be less than 5650 years old.[31] The lower date from core V10–58 is regarded as anomalous; yet even if we discard it, the rest of the operation fails to carry conviction.

Most of the trouble stems from the need of supporters and critics alike to rest their case on a handful of dates. Compare the situation in the study of ocean-floor geology and geophysics, where bold hypotheses are given a due measure of chronological testing. The concept of sea-floor spreading, for example, gains support from the prevalence of a direct relationship between age and depth with increasing distance from the axes of spreading. According to the Vine–Matthews model, mapping the magnetic anomalies imprinted on the floor of the oceans amounts to producing an isochron map;[32] but the final word rests with point age determinations. Admittedly, many such determinations are still based on palaeontology, but the trend towards total reliance on radiometric dating is unmistakable.

Palaeontology stands to gain from the transfer of the dating burden, as the fossil record can then be put to good use in establishing other aspects of the ancient environment which need to be considered in developing plate tectonics. Its interpretation will nevertheless be coloured by the realisation that a spreading ridge system hampers the development of 'layer-cake' stratigraphical relationships and hence that diachronism may prove to be the rule rather than the exception in pelagic sediments.[33]

7 River deposits

It is one of those cases where the art of the reasoner
should be used rather for the sifting of details than
for the acquiring of fresh evidence.
  A. Conan Doyle, *The Memoirs of Sherlock Holmes*

Attempts have been made since Antiquity[1] to trace and interpret
the development of landforms produced by rivers. The subject
figured prominently in the struggle to establish uniformitarianism.
James Hutton drew on the work of streams to illustrate his *Theory
of the Earth*; Lyell, despite a curious blindness to the power of
running water,[2] took pains to establish the age and rate of formation of fluvial features with the help of archaeological remains
and eyewitness accounts. In the ensuing debate between the
fluvialists and the marine erosionists (among them Lyell) the case
for fluvialism owed much to evidence bearing on the effectiveness
of stream action; it may be significant that the origin of planation
surfaces—where such evidence is hard to come by—is the one major
topic still dominated by such doctrinaire debates. Recent years have
witnessed a fresh alignment between exponents of a cyclical view
of fluvial history and those who are more concerned with equilibrium concepts, but there is every hope of a compromise that admits
a past while giving due weight to the present-day situation.

The complementary character of historical studies and direct
observation of fluvial processes is illustrated by Lyell's use of information on the volume of sediment carried by the Mississippi and
on the volume of its delta to compute the time required for its
accumulation. A modern parallel is to be found in attempts to
compare rates of sediment yield in the past, as manifested in dated
alluvial deposits, with values obtained for the existing streams.[3]
Nevertheless, just as some works on fluvial processes pay scant

attention to geological history, so do we meet chronological studies which are solely concerned with subdividing the record into episodes of deposition and erosion.

The discovery of prehistoric artefacts together with extinct faunas in the gravels of the Somme in 1838 initiated a fruitful association between geology and archaeology whose main concern remains chronological. The recognition of Pleistocene glaciation further stimulated the search for alluvial sequences, especially in areas which had lain beyond the ice sheets and glaciers and which therefore promised to supplement the evidence of tills and moraines with other palaeoclimatic indicators. An important component of this work stemmed from the fusion of glacio-eustasy with the concept of base-level control of stream action, as it raised the possibility of linking the fluvial record with that of sea level and hence with the succession of glacial advances and retreats. Add to these the needs of students of soil erosion and of recent crustal deformation, and the literature on alluvial chronology *per se* no longer seems surprisingly large.

Alluvial stratigraphy

Distinctive landforms commonly give identity to bodies of recent sediment—glacial, lacustrine and coastal as well as fluvial—but the resulting items do not always come up to the requirements of 'geologic-climate units' (*see* p. 12). Hence the proposal to introduce the 'morphostratigraphic' unit, which is defined as

a body of rock that is identified primarily from the surface form it displays; it may or may not be distinctive lithologically from contiguous units; it may or may not transgress time throughout its extent.[4]

One of the objections to the proposal is that form alone does not permit the separation of two such superimposed units : the contact between them is buried and thus falls outside the terms of reference of the definition.[5] The American Stratigraphic Code simply states that the constructional morphological character or primary surface of a rock-stratigraphical unit should be subsidiary to the character of the rock itself in defining the unit, and it warns against reliance on 'concepts based on inferred geologic history'.[6] As many workers have found to their cost, a continuous river terrace may be composed of several discrete alluvial units, while an irregular or stepped surface can result from the erosion of a single deposit. There is clearly much to be said for limiting the use of morphological criteria to the preliminary stages of an alluvial study while bearing in mind the possibility that river terraces, piedmont fans and other features redolent of geological inference may in fact not coincide with sedimentary units.

All the above presupposes that the sediments are the primary concern. But there are times (as we shall see below) when the form and composition of the land surface is central to the study. A buried land surface is, of course, no less amenable to age- and form-analysis than an exposed one, always providing it can be traced in exposures supplemented by pits and boreholes (figure 16A).

Furthermore (as the American Code concedes) lithostrati-graphical subdivision is not always plain sailing. A boundary placed within a zone of gradation may be as valid as one which coincides with a sharp lithological contact. Alluvial deposits are sometimes graded and sometimes fragmented into a host of well-defined lenses and thin strata. The solution in such situations may be to identify as formations those bodies of sediment that are separated by clear erosional breaks, and to regard minor disconformities (such as cut-and-fill features) as interruptions in the progress of aggradation. One must then face the fact that the contacts as well as the deposits are diachronous, witness the progressive incision of modern alluvial valleys by headward erosion or by the integration of entrenched reaches, although there is evidence to suggest that the erosion of alluvial deposits is often much more rapid than their deposition.[7]

Whether or not lithological and erosional traits lend themselves to stratigraphical ends, recourse is often had to various other means of differentiating between alluvial bodies into formations and their constituent members. They include fossil soils, intercalated non-alluvial deposits (such as wind-blown sands), artefacts, organic remains, and radiometric dates. Changes in colour and in degree of consolidation may be found equally helpful in the field. In some cases the observed differences between two beds are only of degree and simply facilitate the drawing of boundaries between items whose relative age is derived from the law of superposition. In others (for example where radiocarbon dating is used) there is a net chronological gain.

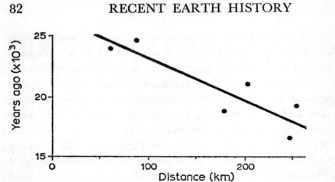

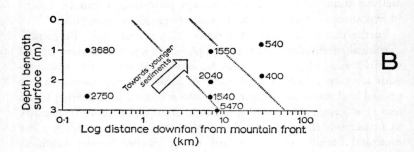

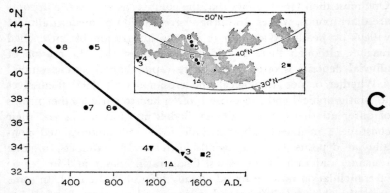

Fig. 16 Time–distance relationships in continental deposits. (A), Plot of ^{14}C age against distance for the base of the Wisconsin loess in Iowa along a traverse from the Missouri River valley to south-central Iowa. Data from Ruhe, 1969, 37–8. (B), ^{14}C dates (error values omitted) related to position within alluvial fan deposits near Biskra, Algeria. After Williams, 1970, figure 6. (C), *Upper,* location of exposures of historical alluvial fill in the Mediterranean area sampled for ^{14}C dating; *lower,* resulting dates plotted against latitude.

Correlation follows conventional stratigraphical procedure. Where the deposits are terraced, however, topographic continuity is an important consideration (as it is in the study of raised beaches), and, despite reiterated warnings of the risks entailed,[8] the relative elevation of adjacent river terraces remains a prominent feature of many alluvial chronologies. Quantitative dates have sometimes dominated correlation, but they are generally appended to stratigraphical units which have been defined in other ways.

Climatic and eustatic frames of reference

Alluvial deposits are subjected to palaeoclimatic interpretation both for its own sake and as an aid to correlation. The two aims are not easily kept separate. To give an example, evidence for increased frost activity during the late Pleistocene has been reported in the wadis draining to the lower Jordan. The deposits in question are poorly dated, but they invite correlation with other indicators of cooler and possibly moister conditions in the eastern Mediterranean, and thereby appear to reinforce the case for such an episode.[9]

Where ample quantitative dates are available, or where the study is confined to a small area which lends itself to lithological and topographic correlation, climatic reconstruction can be deferred until the sequence of depositional and erosional episodes has been established. Unfortunately it is increasingly clear that the factors controlling stream behaviour are too complex for climatic interpretation to be anything but hazardous.[10] S. A. Schumm has shown how rivers draining areas which differ in climate, geology and other characteristics may respond in disparate ways to a major climatic change, and, conversely, how similar responses may be evoked by a wide range of environmental changes.[11] Glaciologists have also shown that the relationship between stream terraces and glacier fluctuations are often complex and that outwash features are a poor guide to glacial history upvalley.[12] Disappointing enough for the palaeoclimatologist, these and allied findings reinforce the need for stratigraphers to guard against homotaxial correlation, as similar successions of similar events do not necessarily arise synchronously. Every valley has a history of its own.

The study of eustatic control of fluvial action has gone through a similar (though not synchronous) cycle of initial enthusiasm followed by chastened reassessment. In its heyday, sea levels served to date and to interpret erosional and depositional sequences far inland and developed in bedrock as well as in the alluvial deposits treated in this chapter. In his classic study of the Isser valley in Algeria (1899), for example, de Lamothe described four Pleistocene terraces each of which could be correlated with a eustatic beach; a similar fourfold succession was recognised by Anderson in the western Atlas (1936). The work of Baulig on the Massif Central of France (1928) was even more ambitious. Besides the vacuum created by the collapse of Suess' eustatic structure,[13] such correlations face the difficulty of demonstrating that sea-level changes will be transmitted upvalley effectively and rapidly enough to be recorded in the alluvial record. What is more, changes in base level (like those due to damming or excavation) can lead to readjustments other than simple aggradation or degradation.[14]

Some workers have wholly rejected a eustatic view of river history; others apply it to the lower reaches in combination with a climatic interpretation of the upper part of the basin. But the phenomenon of 'publication inertia' can mean the perpetuation of existing eustatic correlations. Anderson's Algerian findings are in direct opposition to the field evidence[15] but they find their way into current texts. The Rharbian alluvial formation of Morocco, though shown by radiocarbon dating to be historical in age, is still ascribed to the Neolithic period by virtue of its supposed correspondence with a marine transgression dating from some 6000 years ago, and to some extent through a persistent belief in a North African 'pluvial' period of about the same age.[16]

More usually, the outcome of radiometric dating will be heeded and correlations discarded where necessary; but the gain is the greater if these are replaced by others which are based from the outset on quantitative dates. Radiocarbon dating of molluscs from a silt in central Nebraska has shown that its correlation with the late Kansan Glacial stage had been in error and that the alluvial silts and soils by which it was overlain had formed during the last 10 000 years or so. In the absence of further reference dates, one of the soils was tentatively equated with the 'altithermal soil' of Antevs.[17] While this proposal will doubtless be tested in its turn, it could inadvertently contribute to the continued acceptance of a climatic episode whose status is currently being questioned.[18]

The intention here is not to present strict time-correlation as a panacea. As we have seen, simultaneity can be fortuitous. Quantitative dating merely removes the need to link events by reference to a hypothetical climatic or eustatic mechanism, and enables such postulated mechanisms to be treated as working hypotheses.

Fluvial episodes

Like all point dates, those obtained in alluvial deposits by radiocarbon analysis or archaeological methods cannot be applied to the lithological unit or boundary as a whole without loss of definition. The time-transgressive character of alluvial units is not the only difficulty. A single catchment may be characterised by downcutting at one point and aggradation at another, as when a change in flow conditions is accompanied by a reduction in overall longitudinal profile by incision and extension of the headwaters and building up of the lower channel floor. In such circumstances a single age-determination may give a misleading picture of the conditions prevailing at the time in question.

On the other hand a series of dates may reveal the progress of an aggradational episode or confirm other evidence to the effect that it took place. For example, samples taken at similar depths from Holocene alluvial fans near Biskra, in Algeria, become progressively younger downstream from the mountains (figure 16B). This can be taken to imply that the fall line of alluviation—the mountainward limit of net deposition—has migrated 20–30 km southwards over the period spanned by the dates.[19]

The technique is invalidated by major deviations from aggradation (or downcutting) during the phase under review, as the successive dates no longer apply to a single, cumulative trend. The recent history of the Nile provides an instructive example. Evidence has been found in its middle reaches for a 'siltation phase' which took place between about 20 000 and 10 000 years ago and was followed by a period during which downcutting predominated. The silt is thought to indicate a marked diminution in the Nile's discharge and hence the prevalence of arid and semi-arid conditions in the tropical regions which it drains,[20] a finding which tallies with evidence of aridity in other parts of the tropics at a time when the middle latitudes were being glaciated, and which is clearly of great importance to large-scale palaeoclimatic reconstructions. But the thesis of a single siltation phase has been challenged, and a complex succession of erosional and depositional interludes proposed in its place, raising the possibility that the radiocarbon dates marking the midpoint and the closing stages of deposition in fact refer to two or three separate silt bodies. The problem is still unresolved and has not benefited from the submergence of part of the relevant area under Lake Nasser.[21]

Episodes which are unquestionably shortlived may still command attention because they are represented in widely-separated areas. Quantitative dating can then serve to reinforce time-correlation based on fauna, archaeology, fossil soils and inferred climatic change, and to demonstrate that the observed synchroneities are more than the outcome of using such criteria. In the south-western United States, for example, it has long been claimed that episodes of fluvial aggradation and erosion were broadly contemporaneous over a large area. Using 93 radiocarbon dates whose stratigraphical position was well documented, C. V. Haynes recently showed there was little chronological overlap between the major depositional units recognised in the literature. Nevertheless he took pains to emphasise that, although this points to periods when processes of aggradation were dominant throughout the south-west and others when degradation predominated, it does not follow that erosion and deposition began everywhere at the same time.[22] Such broad correspondences had already been revealed by archaeological and dendrochronological evidence, and were bound to invite inter-pretations involving some kind of climatic change. In the case of the current episode of channel trenching, however, white settle-ment, and particularly the introduction of large herds, appears an equally plausible explanation. There are, of course, those who see overgrazing as little more than a trigger which released an impending erosional trend, and others who regard trenching as the outcome of channel steepening during aggradation. At all events the debate over those parts of the sequence that refer to the period prior to 1870 is unlikely to make much progress in the ab-sence of dependable meteorological, hydrological and vegetational records, let alone where the alluvial chronology itself is in dispute.

A similar impasse characterises the Mediterranean area, where a depositional episode dating from about A.D. 600–1800 is the subject of rival climatic and anthropogenic theories. The episode was observed in Libya and Italy, and subsequently recognised—both in the field and in the literature—in other parts of the basin, and beyond it as far afield as Iran, Germany and the Hoggar Mountains. Both the lithology and morphology of the deposit favour a climatic explanation, as they indicate a change towards more equable discharge régimes than those prevailing prior to and since deposition. The influence of man on geomorphological processes appears to vary widely from place to place, and its history is not always consistent with the view that aggradation stemmed from misuse of the land. The tentative conclusion was that stream behaviour reflected a temporary southward shift of the European depression tracks, and that human intervention had merely accentuated the subsequent resumption of downcutting.[23] Radiocarbon dating has proved useful in confirming that, as suggested by the archaeological evidence, aggradation characterised the entire Mediterranean basin during mediaeval times. In addition, the graph that results from plotting the more dependable dates against lattitude is not inconsistent with the southward lag to be expected from a progressive displacement of the circulation pattern (figure 16C).[24]

Evidently a plot of this nature is inadequate to bear the weight of a grandiose climatic reconstruction, and it is unlikely to convince the sceptic that the deposit represents a single climatic episode. More germane to the general argument, as the radiocarbon samples come from various depths within the deposits rather than from their base or summit (a fault they share with many of the south-western dates mentioned above), they are of limited value for demarcating the period of aggradation. In fact, given the diachronous nature of alluvial erosion and deposition, one can readily see how an insufficiently dense network of dates could prove counterproductive by indicating spurious regional trends.

The difficulties (and costs) of dating are often greatly reduced once attention is shifted to the current episode of channel trenching, as its progress may be recorded in early accounts and perhaps even on maps and photographs. Information on concurrent vegetational and climatic conditions, and on present-day stream behaviour, are sometimes also available. Yet, as Cooke has shown, the problem of 'equifinality' dogs the investigator.[25] However closely he traces the possible links between, say, a wagon road and the arroyo that now follows its course, the contribution made to trenching by runoff concentration eludes precise evaluation.

Local history of sedimentation

Some of the information that does not meet the requirements of stratigraphical subdivision and correlation can be put to good use in investigating changes in the rate at which sedimentation has taken place. It is, of course, possible to derive such rates from existing stratigraphical schemes simply by dividing the volume of sediment represented in each alluvial fill by the time it took to accumulate. On the basis of this method it has been found that rates of sediment production in parts of New Mexico, Wyoming and Iowa during Postglacial times but prior to large-scale human intervention equalled and sometimes exceeded those now prevailing in the same basins.[26] This is a useful corrective to extremist accounts of the effects of modern use on soil erosion, although the possibility cannot be dismissed that the influence of prehistoric man is being consistently underestimated.

Studies of this kind depend for their estimates of current sediment yield on measurements of suspended stream load and of deposition in reservoirs of known age. As those responsible for the estimates fully recognise, it is difficult to specify what fraction of the suspended load would be laid down if aggradation were now in progress. In the middle Rio Grande, where this is the case, about one-third of the total sediment load brought into the valley is being deposited; but to triple estimated past rates of sediment yield in the light of this observation is justified only as a temporary expedient—especially when it is done in basins other than that of the Rio Grande.[27] The difference between localised valley deposition and generalised aggradation throughout the entire drainage network must also be considered when developing comparisons of this sort.

Let us now turn to the rate of sedimentation at individual stations. The need for such measurements might seem most obvious when the age of buried fossils or other remains can only be derived from the thickness of overlying material,[28] but it is always preferable to apply the dating method directly to the object itself or, if the age of burial is in question, to the sediment immediately above it, thus doing away with the need to extrapolate. Where rates are of interest in their own right is in the study of sedimentary environments. As Hudson puts it, 'Although time-scales constructed from an analysis of sedimentary processes are unsatisfactory, these processes are of considerable interest in themselves. Now that we have a radiometric time-scale we can ask questions about them that can be answered objectively'.[29] Needless to say, localised observations will subsequently need to be placed within a broader framework, especially when it comes to interpreting ancient fluvial sediments in the light of their recent counterparts :[30] a delta makes little sense if isolated from its stream or its marine setting. But this, if anything, reinforces the need for adequate dating control, contemporaneity being central to regional integration.

The detail that can emerge from a localised study is illustrated by Larrabee's analysis of a deposit with a maximum thickness of 1 m laid down by the Shenandoah and Potomac rivers at Harpers Ferry between 1861–1864. Over one hundred strata could be identified, and comparison with historical records suggested that two or three floods, each lasting only a few days, were responsible for the sequence.[31] Few sections are as closely dated or well documented as this, but bounding dates for the onset and close of sedimentation, when combined with lithological observations, can still prove valuable. At Qal'at el Hasa, Jordan, an alluvial deposit was being laid down between 10 000 and 2000 years ago, the limiting dates being furnished by archaeological remains and supported by a radiocarbon age of 3950 years for a point 1 m below the surface of a deposit 4 m thick. The nature of the constituent strata reflected flood deposition; whether this was seasonal or not, it suggests that conditions suited to flood irrigation locally prevailed at a time when the region as a whole was witnessing some of the most crucial experiments in the history of agriculture.[32] A similar valley-floor environment is indicated by the 'high terrace' deposit in the Tesuque valley of New Mexico.[33] It is noteworthy that the potential for flood-farming in both reaches was drastically reduced by channel incision. The ultimate causes of this incision cannot be inferred from the evidence at hand, but their investigation can be shelved with little compunction.

Such environmental conjectures can be amplified by detailed analysis of the sediments. By comparing the mechanical composition of two successive dated fills at the same section where the younger deposit can be shown to consist principally of sediment derived from the older, some idea can be gained of the régime of flow that characterised the second aggradational phase. The procedure does away with the need to postulate a normal distribution or any other hypothetical curve for the 'ideal' granulometric curve, the parent deposit providing the standard against which the daughter sediment can be compared. Applied to the surface texture of the constituent grains as revealed by electron microscopy, this comparative approach may also throw light on the chemical conditions prevailing locally during redeposition of the particles and on the rate at which the textures are modified.[34]

Site environments

The counterpart to the archaeological dating of alluvial deposits is the geological dating of archaeological sites. Although most common in cave studies (*see* chapter 8), it is also practised with the help of alluvial deposits. The risks that attend the method are brought out by the history of the Keilor skull from Australia. The skull was found in a river deposit which could be correlated with a fossil beach, and this in turn was correlated with the Main Monastirian sea level of the Mediterranean. As this had been dated by reference to the Milankovitch (solar radiation) curve to about 150 000 years ago, a similar age was ascribed to the skull. Later, radiocarbon dating of an alluvial deposit to which the source bed could be correlated showed the skull to be some 5000 years old.[35] A radiocarbon date of 18 000 years has since been obtained for a lens of charcoal and bones within the alluvial deposit, but the lens is not definitely associated with the skull.[36]

K. W. Butzer has dealt in some detail with the study of alluvial sites, and in particular with their environmental analysis in terms of human occupation. Although some sites to which he refers have been interpreted (and dated) chiefly by reference to regional palaeo-climatic sequences, many of them illustrate the value of concentrating on the local situation—without wholly ignoring the local setting—when interpreting cultural and faunal remains.[37] Work along similar lines in Epirus (Greece) has been directed chiefly at tracing the history of soil erosion and valley filling during pre-historic and historical times, the aim being to complement the faunal, archaeological and documentary record with information on changes in the area's economic potential. The closeness with which the sediments can be dated is limited by the need to depend on archaeological methods, but this drawback has been minimised by radiocarbon dating of the various artefact assemblages repre-sented both in the excavated sites and in the alluvial fills; as regards the last 2000 years, classical and later sites provide sherds and other debris which enables fluvial history to be traced in some detail. Climatic reconstruction has been restricted largely to changes in the position of the permanent snowline, as this bears on the territory available for exploitation at the various sites at different times of the year; it is worth adding that the results support the site evidence for seasonal movements in Palaeolithic times analogous to those now practised by transhumant herders.[38]

Preliminary work in Mexico has already shown that the dating of alluvial fills in the vicinity of some of its major archaeological sites is likely to prove rewarding. For example, the alluvial infill of the Salado river near Tehuacan is divisible into two units (figure 17), of which the older ceased undergoing deposition not later than about 8000 B.C. and the younger dates from the first millennium A.D.[39] Without indulging in an unpalatable degree of determinism one may suppose that the sequence played some part in the valley's complex agricultural and faunal history.[40]

The chronological study of alluvial sites may lead on to wider issues. According to R. L. Raikes, Mohenjo-daro was buried by mud which accumulated in response to localised uplift in the lower Indus. Conversely, he explains the 'otherwise almost inexplicable siting' of Neolithic Beidha, in Jordan, to erosion of the adjacent alluvial terrain following rejuvenation of the valley occasioned by faulting or tilting downstream of the site.[41] Besides the stress on local fluvial history, we must welcome the diversion from palaeo-climatic to tectonic interpretation of the evidence.

But, as Lyell pointed out, 'In all calculations referring to the growth of alluvial deposits, or to the effects of aqueous denudation,

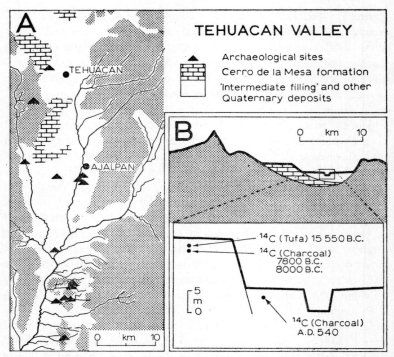

Fig. 17   Dated alluvial fills in the Tehuacan Valley, Mexico. (A and B, *upper*), Geological map and section of the valley at Ajalpan. Modified after Brunet, 1967, figures 41 and 44. (B, *lower*), Subdivision of 'Intermediate filling' into two distinct deposits (dates after Vita-Finzi, 1970b).

our chief difficulty in geology arises from our inability to measure correctly the accompanying movements of the land'.[42] Movements which postdate deposition are more readily determined provided the deformation can be isolated from similar effects due to fluvial processes. This is the case at Gafsa, in Tunisia, where folded gravel beds have yielded Acheulian artefacts and are overlain by undeformed alluvium which contains Mousterian material. It has been suggested that the Acheulian artefacts are confined to a superficial calcareous crust, and hence cannot provide a maximum date for the folding,[43] but many workers adhere to the original interpretation.

The rupture of fluvial deposits is of course easily observed even when slight. Both transcurrent and vertical faulting of alluvial terraces have been reported from New Zealand, with displacements of as much as 60 m since the last glacial episode in the area.[44] Faulting of recent alluvial fans along the Dead Sea Rift is also well documented; in one instance the horizontal movement amounts to 150 m and is thought to have occurred within the last 20 000 years.[45] Quantitative dating of such faulted beds[46] makes it possible to convert displacement into mean annual rates without gratuitous conjecture; given an adequate number of age determinations, the role of stratigraphy is confined to delimiting what were originally geometrically continuous units.

8 Cave infills

> The old partisan Dubovoy . . . even developed in
> jail the theory that the Purge was the result of an
> increase in the number of sunspots.
>
> Robert Conquest, *The Great Purge*

Caves and rock shelters are prominent in the literature of Quaternary palaeontology and prehistory, as the protection they afford from the elements both invites occupation and favours the preservation of the concomitant skeletal remains, hearths and tools. In fact, many of the classic subdivisions of the Middle and Upper Palaeolithic of Europe were defined on the basis of such material, notably the Mousterian (after the shelter of Le Moustier), the Aurignacian (after the cave of Aurignac), and the Magdalenian (after the shelter of La Madeleine).

The excavation of caves* may be purely exploratory, or it may have as its aim the elucidation of some specific aspect of human, faunal or environmental history. The techniques employed vary from worker to worker, although some national 'traditions' have been recognised. Nevertheless the subject gains unity from the widespread use of climatic change as a means of dating and correlating cave sites.

Climatic history

The processes of sedimentation within a cave are influenced by its form, position, aspect and history of occupation; deposition itself affects conditions within the cave if only by progressively raising its floor. The proportion of sediment derived from outside the cave

---

* For the remainder of this chapter 'caves' will be used for all such features ranging from shallow overhangs to deep caverns.

is no less variable. As a result, two caves which lie close together, and indeed different portions of the same cave, may differ markedly in character at any one time. Palaeoclimatic reconstruction works on the assumption that pronounced environmental changes will eclipse local anomalies and leave recognisable traces. In Périgord, for example, various techniques have been employed to evaluate the intensity of frost shattering and of chemical processes acting during the late Würm and, in some cases, as far back as the latter part of the Penultimate (Riss) Glaciation.[1]

Some workers prefer to focus their attention on components of the cave infill which are of external origin and which presumably reflect local conditions undistorted by the idiosyncracies of the cave environment. C. K. Brain terms them 'phase 2 deposits'; his reconstruction of rainfall changes at the Cave of Hearths in the Transvaal is based primarily (though by no means solely) on the analysis of siliceous sand grains which came to be incorporated within the dolomite breccias of the cave.[2] Again, studies which rely mainly on faunal or vegetational evidence of climatic oscillations implicitly view the cave as little more than a repository for a sample of the local population : take, for example, the attempt by D. M. A. Bate in Palestine to trace fluctuations in climatic humidity by plotting the relative frequency of fallow deer and gazelle remains at different levels in the cave deposits of Mount Carmel.[3] But it is more usual for climatic information to be sought from both internal and external sources, and for the evidence of sedimentology to be coupled with that of palynology, zoology, isotopic palaeotemperature analysis and other relevant disciplines.

The agreement between diverse climatic indicators is sometimes convincing, as in the case of beds D and E in the Weinberg caves near Mauern, in Germany, where a limestone rubble indicative of cold conditions has yielded an 'extremely glacial' fauna.[4] If the results fail to tally, the explanation may be found in intrusion of material from a higher horizon, say through the agency of burrowing animals, in the fact that the conflicting results were obtained for sequences from different parts of the same cave, or in the erroneous interpretation of one of the indicators. The ambiguity of much of the evidence is, of course, a constant problem. Phosphatic cave beds are sometimes ascribed to arid conditions, but it has been shown that they form in cold as well as in warm environ-

ments, and that the phosphate content of bear caves is affected
by changes in the ratio of sedimentation intensity to the number
of dead animals.[5] Angular fragments can be produced not only by
frost activity but also by mechanical adjustment of the cave void,
crustal wedging, hydration, and other mechanisms. Calcareous
deposits are difficult to interpret with conviction.[6] The environmental requirements of many organisms are equally uncertain; if
their modern equivalents are taken as a guide there is ample scope
for the overlap of species with conflicting preferences; hence the
frequent reminders that the fauna and flora need to be considered
as a whole before any climatic inferences are drawn.

Dating the record

By slotting the climatic history of the cave into a time-calibrated
regional chronology, the age of its constituent horizons can be
divined. This procedure in turn yields further details of the regional
succession, as may be seen from the literature devoted to the caves
of the Mediterranean area.[7] Where quantitative dates are sought,
correlation is sometimes practised over long distances; thus, the
lower 40 per cent of the succession in the Haua Fteah (Libya)
yielded isotopic temperature readings which could be matched with
faunal and isotopic curves derived from cores taken from the floor
of the eastern Mediterranean, and these were in turn correlated
with an Atlantic core for which radiometric dates were available.[8]
The choice of 'standard' scheme is important, especially when
dealing with the early Quaternary, as the alternative results can
diverge appreciably. The remains of Peking Man at Choukoutien
were first placed in an interglacial by virtue of the associated fauna,
but pollen studies subsequently showed them to be of Mindel-II
age.[9] According to one chronology the fossils would consequently
date from about 300 000 years ago and according to another from
over 1 000 000 years ago.[10]

The possibility that the correlated sequences are out of phase by
one glacial/interglacial cycle or more is again of greater moment
the coarser the subdivisions of the dated reference column. Where
the cave or shelter occupies a littoral position, the presence of beach
material or of fossil dunes within the infill may provide a helpful
reference point. Many of the caves on the Mediterranean coast

have thus been dated by reference to features which developed
within them during the Last Interglacial transgression; the detailed
marine sequence worked out for Atlantic Morocco has enabled
human remains found in its coastal sites to be dated as far back
as the Günz–Mindel Interglacial (that is, to the first of the inter-
glacials in the classic Alpine scheme).[11] But if—as in the above
examples—the reference point is also dated by climatic correlation
it may be less than secure. At Gamble's Cave II, in Kenya, lake
beaches formerly equated with stages within the Würm are now
thought to be no more than 10 000 years old.[12] Quite apart from its
archaeological implications, this reappraisal has not benefited the
glacial = pluvial equation, of which the Gamblian Pluvial long
served as an illustration.

On occasion the basis of correlation is purely biostratigraphical.
Individual mammalian species have served as index fossils, among
them the hyaena *Crocuta crocuta*, which is thought to have spread
rapidly through Europe during the Middle Pleistocene and which
has helped to establish cave deposits as being no older than this.[13]
But it is generally found preferable to rely on faunal assemblages
which characterised particular periods, such as the 'land-mammal
ages' of North America or the 'faunal spans' proposed for Africa.[14]
The resulting correlations are as prone to reassessment as those
that hinge on climate : for instance, the fauna yielded by a fissure
in Podlesice, in southern Poland, has been placed successively in
the Middle Pleistocene, the Villafranchian and the Pliocene.[15]
Moreover, the faunal ages and spans are generally too broad and
ill-defined to be of much value to the stratigrapher.[16] At all events
there appears to be a strong temptation to equate them with subdivisions of the Quaternary derived from palaeoclimatic evidence,[17]
thus reducing the independence of any contribution they might
make to chronology.

The archaeological dating of cave deposits conforms to usual
practice. The fact that it yields units finer than those based on
fauna means, among other things, that it is used to check palaeontological correlations. For example, a Mousterian site at Krapina,
in Yugoslavia, had been ascribed to the Last Interglacial because
it yielded the remains of Merck's rhinoceros, reputedly a tropical
species. This conclusion was later challenged on the grounds that
the same rhinoceros occurs in Spain up to the Aurignacian.[18] Yet

radiocarbon dating suggests that, at least within France, the occupation of some Mousterian and Aurignacian sites overlapped.[19] More generally, it is difficult to justify the assumption implicit in some current studies that 'cultures' were time-parallel. At all events, the search for a 'cultural-stratigraphic' terminology consistent with geological usage[20] presages continued reliance on homotaxial correlation.

The status accorded to radiometric methods varies widely even though the preservation of organic material and hearths within caves invites radiocarbon dating. Thus a recent site report explains that potassium–argon ages had to be obtained solely because there

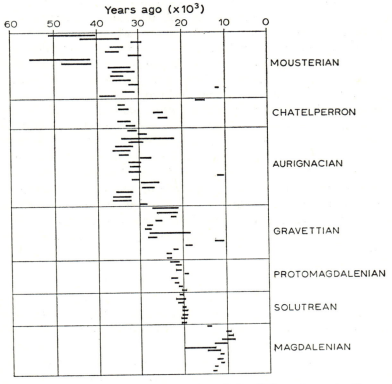

Fig. 18 Radiocarbon dates for various Palaeolithic cultures in France. Bars indicate range of three standard deviations. Modified after Coles and Higgs, 1969, figure 5.

was no suitable faunal material. Discrepancies between radiometric
dates and those stemming from climatic correlation have raised
doubts in the dating methods : the fact that the radiocarbon determinations from the upper part of the Haua Fteah are 'shorter'
than the estimates derived from correlation with Atlantic cores is
thought to indicate either that the dates for the cave are all too
young because of contamination or that the dates for the cores are
systematically biased towards high readings.[21] Again, the overlap
indicated by radiocarbon dating between successive Upper Palaeolithic industries (as well as between the Aurignacian and the Mousterian) in France (figure 18) has been disputed in terms both of
inadequate stratigraphical evidence and of the 'relative coarseness'
of the radiocarbon method.[22] Dates which are grossly contrary to
expectation may be dismissed as 'still unreliable', 'unacceptable',
'difficult to believe', and sometimes more trenchantly.

The relevant dates are indeed few and often suspect, although
a lack of enthusiasm for radiometric age determination may also
stem from the low priority hitherto given to the issue of contemporaneity in palaeontology and prehistory. But the situation is
changing. Take the suggestion that, rather than use *Crocuta crocuta*
as a dating fossil, we should try to find it in datable deposits in
order to trace its spread and early evolution;[23] or the use of dates
determined directly on the remains of Woolly Mammoth in Siberia
to trace oscillations in its size through time;[24] or the emphasis on
radiocarbon dating in assessing the role of man in late Pleistocene
mammalian extinction.[25] And at least one prehistorian already finds
it 'no longer possible to think of the correlation of cave deposits
on a geological time scale. A thousand years or less is of some
importance.'[26]

The uniqueness of caves

The above adds little to the argument of earlier chapters, and
is in fact intended to demonstrate that the stratigraphical study of
caves raises all the usual problems. But it is right to treat cave
deposits in the same way as river sections, ocean cores, beach
profiles and other means of sampling past environments?

A cave is attractive to its prospective occupants largely because
it departs from the conditions that prevail in its vicinity. Besides

creating difficulties for the palaeoclimatologist, this can mean that the faunal and archaeological sample in the cave will be biased towards those species that stand to benefit from the cave environment, and the food debris or artefacts associated with their presence. In contrast, whether or not an ocean core or quarry face actually yields a representative palaeontological sample, the working assumption (backed by apposite sampling procedures) is that it will. Admittedly, many other kinds of site are environmentally anomalous, most obviously in the case of oases, but this hardly justifies disregarding the problem where caves are concerned.

What is already an unrepresentative sample may be further distorted by stratigraphical demarcation. The literature contains many warnings against assuming that breaks in the faunal or archaeological record coincide with the boundaries of sedimentary units, as they could just as well occur in the course of a depositional phase. To this may be added the rider that cave strata, like their external counterparts, can easily deviate from time-parallelism. Unless a large number of radiometric dates is to be had, only the presence of an unmistakable living-floor will indicate that different parts of the cave were occupied simultaneously. Some workers take pains to record the position of excavated material within each layer by reference to an arbitrary grid, though fortunately without making the claim that horizontality is an infallible guide to contemporaneity; others 'have approached the limits of micro-stratigraphic control in which complete and fully documented artefact assemblages are segregated according to the minimal sedimentary components that can be discerned in archeological *couches*'.[27] Thus the fauna or 'industry' will be defined on the basis of association either within a *couche* or in three-dimensional space.

The risk that the association does not represent reality is increased when deposition involved external agencies operating concurrently with occupation. The Last Interglacial fauna in the Eastern Torrs Quarry cave near Plymouth was laid down by running water, and is therefore no worse a guide to the local population than a river fill. On the other hand the lower part of the cave of Tabun, on Mount Carmel, gains most of its stratigraphical complexity from the cross-bedding of dune sands blown into its mouth, and it is difficult to accept the industrial units that have been recognised within this material as a faithful reflection of cultural progress. The deposits at La Micoque, a shelter in the Dordogne, include much scree material whose accumulation was probably accompanied by the reworking of material from higher up the slope. This may help to explain why the six industries that were differentiated stratigraphically show a 'remarkable overall uniformity' (as well as some strange anomalies) and why La Micoque remains the only site in the Dordogne yielding the industry named after it.[28]

In his classic paper on 'bone caverns', published in 1833, Tournal outlined some of the many ways in which fossils and deposits could be introduced into caves, and stressed the need for evaluating each site on its own merits.[29] His concern was to refute the view then current that silts and gravel beds within caves had been laid down by the Deluge; it appears that the point needs restating with regard to other aspects of stratigraphical interpretation.

Cave catchments

The stable conditions that prevail in deep caverns nevertheless offer certain advantages in the study of ecological, mineralogical and other processes.[30] Here, and in seeking to elucidate the sedimentary history of a single cave, a firm chronological framework is needed. In a recent study of sedimentation rates, the cave strata were dated by comparing the inclination and declination of their remanent magnetism with a regional archaeomagnetic time scale itself calibrated mainly by tree-ring methods,[31] and one may look forward to the application of more direct dating techniques to this and other aspects of cave environments.

The corresponding disadvantages of a cave situation can also be put to good use, though again on condition that interpretation of the evidence is not dominated by the needs of chronology. Kurtén has shown how the very large numbers of cave bear (*Ursus spelaeus*) represented in some caves—up to 30 000 individuals—may result from very intensive 'sampling' during winter, the season of peak mortality. The standing population may thus have been quite small. By applying suitable analytical techniques one can glean a good deal of information from the skeletal material about the mortality of different age groups and the dynamics of the population.[32] This kind of analysis is of interest in many fields of ecological research and notably in attempts to isolate the factors responsible for selection or extinction during the specified periods.

As we have seen, the cave infill will supply some of the evidence relating to external conditions that is usually required in interpreting the faunal (or mineralogical) evidence. It is sometimes claimed that caves, unlike bogs, yield pollen grains derived from the immediate vicinity as well as that carried in on the fur of animals and on the feet and clothing of men,[33] but long-distance transport cannot easily be discounted. A similar claim is made for the analysis of the rodent remains introduced by owls, on the grounds that the slightest of climatic changes will be reflected in the 'rodent-spectrum' through shifts in the range of the vegetational environments exploited by the owls.[34] The method has much to commend it, granted the risk of bias introduced by a preference for certain kinds of prey. But once again the outcome is to give most prominence to environmental reconstruction, as when the various species of rodent accumulated in the caves of Mount Carmel by the Barn Owl (*Tyto alba*) were classed according to their preferred habitats—'rock-dwellers', for example, or 'swampy steppe dwellers' —so that changes in their relative importance over the last 120 000 years could be interpreted in climatic terms.[35]

As it happens, the Mount Carmel rodent record is adequately explained, without the need to invoke climatic fluctuations, by the erosional and depositional changes undergone by the owl territories and recorded in the local alluvial sequences. The application of 'site catchment analysis' to the problem of Palaeolithic occupation of the area has proved equally informative. Site exploitation territories were defined by radii representing two hours' walk from the site; the resulting boundaries took account of variations in topography and lay well within the 10 km limit reported from gathering and collecting economies of the present day. The territories were then subdivided into terrain categories flexible enough for the purpose and amenable to correction for physiographic changes since Palaeolithic times. Once the resource potential of each site territory had been assessed, hypotheses could be formulated regarding the character of the economy they supported and the degree of complementarity between the sites. Hence, although the initial stress was on the individuality of the territories, the outcome had a bearing on regional patterns of resource exploitation.[36]

Working from the setting to the site (rather than the converse) requires that the time relationships between the internal and external sequences become unambiguous. The ideal is approached by the cave of Kastritsa, in Greece, whose territory consists largely of lacustrine deposits laid down during a period of high lake level represented by a set of fossil beaches within the cave deposits.[37] Archaeological dating of the physiographic evidence may be no less satisfactory provided that the reference material is obtained from the cave under investigation. Generalised climatic and eustatic sequences are always to be treated with caution, witness the fact that the seasonal marshes whose presence between Mount Carmel and the sea is essential to the understanding of the area's Palaeo-lithic economy owe their disappearance largely to artificial drainage at the hands first of Roman and then of Zionist settlers.

Independent chronometric dating of the sequences will clearly facilitate the task of time-correlation; where it becomes essential is in the testing of hypothetical regional patterns, and in tracing the long-term changes ephemerally represented in the cave infills. Archaeologists have recently been reminded that they are well fitted to develop techniques 'for measuring variations in the demographic and behavioral characteristics' of former populations over long periods of time and in relationship to their ecosystems.[38] The essential preliminary remains that of keeping the various strands of the problem chronologically separate until it is safe to intertwine them.

9 Cases and laws

> I just report the facts and any data—such as the gins
> and the weather—that seems to have a bearing. It's
> for the others to draw the conclusions.
>
> Grahame Greene, *Travels with my Aunt*

The examples we have now considered will have done their job
if the reader should henceforth regard it as odd for an investigation
into earth history not to display some evidence of a strenuous
attempt at time-reckoning. That stratigraphy can often yield the
desired answers with little or no help from quantitative dating is
manifest in the achievements of petroleum geology, hydrogeology
and civil engineering. But where time is the primary concern we
cannot afford to skimp on dates.

The need to compromise will of course remain. For, as we have
seen, one does not often find datable material exactly where one
wants it, and there is poor consolation in being reminded that many
of the stratigrapher's beds are unfossiliferous and that many of his
depositional sequences include hiatuses. The danger lies in allowing
compromise to eclipse the extreme positions it was designed pro-
visionally to bridge. In illustration let me quote a critic of the view-
point expressed in these pages.[1] 'Although something of a revolution
in stratigraphic thought has indeed been triggered off by the
appearance of numerical dates (he writes), their impact on stratigraphy up to now has been small because few of them can be given
stratigraphic meaning.' But (he goes on), all is well, as stratigraphers
'are adept in the business of juggling with two sets of chronological
data simultaneously'. What has been overlooked is that the act
becomes gratuitous once stratigraphy ceases to be the end in view.

It seems prudent to see how well the chronometric ball behaves
before giving it one's sole attention. But a half-hearted attitude to

dating is unlikely to improve its performance. That every geology department, and perhaps also every institute of archaeology, should have its own radiocarbon laboratory, and that other kinds of radio-metric dating facilities should be freely available, seems an extravagant request; yet the laws of supply and demand—as makers of scanning electron microscopes are well aware—apply to scientific instruments no less than to other consumer durables. At present dating is something of a luxury, and so it will remain until it comes to be regarded as an essential.

Time-tables

Even so, many will agree with the eminent archaeologist who advised his colleagues not to allow chronology to monopolise their discipline. To be sure, without dates 'in hard figures' it was impossible to reconstruct the causative factors of human progress or the fluctuating tempo of human achievement. But, as he put it, enough of preparing time-tables : 'let us now have some trains'.[2]

This is admirable, provided the time-tables are drawn up before the trains start running. And it is perhaps nearer the mark to speak of recording the actual times of arrival and departure as accurately as the available clocks will allow. Chronology can then be pushed into the background, where it belongs. Compare the situation that prevails when the stratigrapher tries to match a local sequence with an existing standard : chronology is at the forefront, and its potential applications take second place.

Enough has been said in earlier chapters to indict climatic correlation as a prominent example of what could be called philatelic earth history if greater value were accorded to imperfect specimens. For, even when dependence is placed on fossil soils and other features which are thought to deviate little from time-parallelism, the retention of climatic change as the basis of time-stratigraphical subdivision has as it corollary the search for 'standard' classifications, whether regional or provincial. The extreme to which the field investigator may then be driven in order to explain away the 'missing' portions of his record are well illustrated by the Mediterranean literature : on the assumption that every yodel in the Alps had its echo on the coast, pebble bands are equated with glacial episodes, truant beds are eroded away, and the uplift of mountains is delayed to justify the absence of glacial features.[3] Equally instructive is the case of the Alleröd, a warm episode which has been identified in the record of much of western Europe and which was apparently followed by renewed cooling about 11 000 years ago. Granted its global incidence, the absence of a lithostratigraphical equivalent in North America can be explained as a failure by the Alleröd to leave a clear imprint on the landscape;[4] but why not accept that it was restricted to Europe and then see how far this can be explained in terms of the local meteorological situation?[5] 'If absence can prove the same thing as presence, anything goes.'[6]

These and kindred comments could be dismissed as the flogging of dead horses, at least as regards East Africa where the use of climate as the primary basis of stratigraphy has been authoritatively censured.[7] But the horse's death is not common knowledge;[8] more important, what has been proposed in place of climate is the correlation of local lithostratigraphical sequences with the help of faunas, cultural materials, geomorphology and the other stratigraphical tools coupled, where possible, with isotopic and palaeomagnetic dating. In brief, palaeoclimatic correlation is a symptom and not the cause of the search for regional uniformities implicit in any kind of correlation.

Where a standard sequence does appear to be represented in the area under review, a massive dose of dating will naturally help to decide the merits of the case. The operation can prove especially rewarding where it is decreed to failure by scholastic correlation : radiocarbon dates recently obtained for British deposits hitherto regarded as too old to be tackled by the method suggest, among other things, that features thought to have formed between 265 000 and 200 000 years ago may be no more than 44 000 years old.[9] But the real point at issue is whether the concept of a standard succession is a fruitful one.

'The factual burden of science', writes Medawar, 'varies inversely with its degree of maturity. As a science advances, particular facts are comprehended within, and therefore in a sense annihilated by, general statements of steadily increasing explanatory power and compass—whereupon the facts need no longer be known explicitly, that is, spelled out and kept in mind.'[10] Whatever the merits of correlation, its outcome conforms to the above criteria of scientific maturity only as regards the annihilation of particular facts; that type sequences are superior in explanatory power to detailed accounts of local anomalies remains to be demonstrated. It is perhaps no coincidence that, in his celebrated review of *The Phenomenon of Man*, Medawar should (twice) remark that Teilhard de Chardin practised an intellectually 'unexacting kind of science'.[11] Although there was little alternative open to him, Teilhard pursued a brand of palaeontology which did not balk at applying the nomenclature of France to the prehistory of China.[12]

The contradictory properties often ascribed to type sections is summed up in the assertion that 'If the stratigraphical code is to attain objectivity the type section must possess the quality of uniqueness'.[13] While we may speak of characteristic or typical successions, investing them with general validity invites fallacious interpolation elsewhere for the sake of uniformity. J. A. Wilson has urged us to 'recognise type sections for what they really are, samples',[14] and in so doing has demolished the case for having them at all.

Nonhistorical earth history

In seeking an alternative and more productive mode of generalisation it is helpful to glance again at the antecedents of present-day practice. Kuhn describes historical geology as it was before Hutton (1726–97) as a field which was something less than a science in lacking a universally accepted body of beliefs, and he cites Lyell's *Principles of Geology* as a work which, by supplying the missing paradigm, opened the doors to the puzzle-solving that characterises 'normal science'.[15] What is it about post-Lyellian geology that unites the scholars against the cranks? Most would include among its articles of faith the law of superposition of strata, the tenet that beds may be correlated by means of their fossil content, and the Principle of Uniformitarianism. Only the last of the three can be attributed to Lyell, and this only to the extent that he succeeded in universalising Hutton's central idea. At all events it is uniformitarianism which Kuhn takes as the prerequisite for admission into the geological fraternity.

Lyell argued that geological interpretation could not proceed without confidence in 'the permanency of the laws of nature.'[16]

In his reforming zeal, he went on to explain the past in terms of 'existing causes' operating at rates currently to be observed. In the minds of some of Lyell's disciples (and critics) the two concepts became fused. S. J. Gould has succeeded in separating them again by terming one 'methodological' and the other 'substantive' uniformitarianism.[17] We may argue whether Gould and Lyell before him were right in equating a belief in the invariance of natural laws with a warrant for inductive reasoning, for this would make geology the one science to practice what Bacon preached.[18] What the belief undoubtedly does represent is a licence to perform science. There is therefore nothing particularly geological about this sort of uniformitarianism. And, now that the revolution heralded by Hutton and Lyell is being realised because the aeons they bestowed on earth history can be reduced to measured quantities, so will geologists find less and less scope for a paradigm of their own, a loss which doubtless only historians of science will mourn.

What of substantive uniformitarianism? This is now dismissed as 'false and stifling to hypothesis formation', just as catastrophism was seen by Lyell as a dogma 'calculated to foster indolence, and to blunt the keen edge of curiosity'. But, by adopting time as the sole basis of chronology, and thus allowing each sequence to speak for itself, we do away with all such rules of thumb and at the same time clear the ground for generalisation of another kind, namely that inherent in explanatory statements. For, if the events considered by 'historical sciences' are rarer, more complex, and therefore less amenable to explanation in terms of scientific laws than the events treated by 'nonhistorical sciences', it is not because there is any logical difference between the two sets of events. The difference is empirical, and stems from the relative difficulty with which they may be observed or subjected to experimentation, as well as from our limitations as observers.[19] Compare, for example, the situation today with that prevailing forty years ago, when discussion of a theory bearing on the relationships between volcanic activity and mountain building was dismissed as unprofitable because the available chronologies were too crude.[20]

The temptation persists to exclude certain aspects of earth history from a chronometric scheme albeit on grounds other than of ignorance. In geomorphology, for example, a 'timeless' (as opposed to a 'timebound') approach is held to be applicable where the relationships between stresses acting on the earth's surface and the resistance of earth materials to them are established virtually instantaneously[21] and therefore benefit little from historical analysis. An interesting variant is found in the recognition of time spans of different order, 'cyclic time' being the span required to encompass an erosion cycle, 'graded time' being one during which a graded condition or dynamic equilibrium prevails (say 1000 years), and a 'steady time' span being sufficient for a true steady state to be established (a year or less).[22]

Such schemes bring the metaphor of the arrow in flight virtually instantaneously to mind. Moreover, one wonders how the problem can be launched *a priori* into a timeless void. It is certainly helpful to be reminded that 'effects' can play the role of 'causes' (and *vice versa*) once the temporal frame of reference is changed : for example, the nature of flow within a river channel will be influential in modifying its geometry, whereas, in the short term, channel form has the upper hand. But the proposed units warrant acceptance only if their boundaries coincide with discontinuities analogous to those introduced by scale in the physical conditions that govern life.[23] Take, for example, the time ranges recognised by some geophysicists when dealing with the behaviour of the earth's crust and mantle in rheological terms, namely 'short' (up to four hours, with the material behaving as an elastic solid), 'intermediate' (four hours to 15 000 years, with behaviour comparable to that of a Kelvin body) and 'long' (over 15 000 years, Bingham body).[24] And even here the categories can be considered only after the duration and rate of application of the stress in question has been determined.

In considering issues of this kind it is possible to envisage the day when dating will enable us to disregard age. As Bertrand Russell has remarked, 'no scientific law involves the time as an argument, unless, of course, it is given in an integrated form, in which case *lapse* of time, though not absolute time, may appear in our formulae'.[25] An emphasis on rates and durations evidently does not imply the neglect of sequence and progressive change. To speak of evolution solely in terms of tempo would be a mockery of the concept. Again, the historian may object that the processes of nature are 'sequences of mere events', whereas those considered by the archaeologist and the historian 'have an inner side, consisting of processes of thought'; hence 'the palaeontologist, arranging his fossils in a time-series . . . is thinking in a way which can at most be described as quasi-historical'.[26] Be that as it may, there is an implicit assumption here that chronology can be taken for granted, a state to which earth history can still only aspire.

We may conclude by asking whether there is anything to be gained from putting together again the millennia dismembered for the sake of illustration. The outcome need not be a denial of the principle that problems are preferable to periods, provided the aim is seen to be that of providing a meeting place for workers in the various fields that share an interest in the recent past. Any other slice of earth history would be eligible once it lent itself to a similar level of chronometric treatment.

The immediate past does, of course, have the attraction of serving as a bridge between present-day processes and their fossilised counterparts. Several themes in physical geology suffer from falling between the stools of historical geology and geomorphology. To judge from the literature, the distinction drawn between palaeontology and genetics, or archaeology and anthropology, is sometimes equally deleterious even if largely administrative. But the adoption of a single, quantitative scale of time measurement does more than add coherence to the history of various sets of biological or physico-chemical phenomena : it facilitates the study of their interrelationships. Soon after the radiocarbon method had been developed, F. Johnson observed that it would enable chronological problems which were interdisciplinary in scope to be solved by a direct approach : the method was world-wide in its potential application and, as all its dates depended on a single technique which was independent of geological, archaeological or analogous interpretations, chronologies in several fields could now be collated.[27] All this is true, *a fortiori*, of chronometric dates derived from historical as well as radiometric methods.

But one must beware of formalising such spheres of interest. In 1883, Arthur Evans remarked 'to confine a Professorship of Archaeology to classical times seems to me as reasonable as to create a Chair of "Insular Geography" or "Mesozoic Geology". . . .'[28] Earth history is already sufficiently compartmented. Our real aim should be to ensure that neither the Moon nor Mars come to be compared to bound volumes of history. As regards the Moon it is probably too late.

Notes

1 INTRODUCTION

1 Zeuner, 1958, 5. Zeuner saw as the chief objective of geochronology 'the development of time-scales in years which extend back into the distant past beyond the historical calendar' (*ibid.*, 5). Later in the book he narrowed the definition down to 'dating in terms of years those periods to which the historical calendar does not apply' (*ibid.*, 5). Most writers include both historical and prehistoric times within the scope of geochronology (for example, Leopold *et al.*, 1964; Smiley, 1955).
2 It is true that the radiocarbon and potassium–argon methods are not mentioned until the last page of his *The Pleistocene Period* (Zeuner, 1959) but they figure prominently in the second and later editions of *Dating the Past*.
3 Flint, 1971, 395.
4 Faul, 1966, 52.
5 Geochronometry is sometimes used to denote the dating of rocks pure and simple.
6 De Cserna, 1972, 194.

2 STRATIGRAPHICAL EARTH HISTORY

1 Toulmin and Goodfield, 1967, 202.
2 *See* E. Gibbon, *The Decline and Fall of the Roman Empire*, chapter 15, n. 96.
3 Adams, 1954, 3.
4 Hedberg, 1961b.
5 American Commission on Stratigraphic Nomenclature, 1961.
6 Krumbein and Sloss, 1963, 332. Rock-stratigraphical units can take the guise of parastratigraphical units, where correlation depends on attributes (such as heavy mineral content) yielding boundaries which do not necessarily correspond with those conventionally established at the columns being correlated (*ibid.*, 333–5).
7 Shaw, 1964, 82.
8 Arkell, 1956.
9 Donovan, 1966, 19; Interdepartmental Stratigraphic Committee of the USSR, 1966.
10 The date suggests that the citation comes from *Principles of Geology*.
11 Huxley, 1862.
12 Compare the statement by Darwin (*Origin of Species*, chapter 10) that 'Unfortunately we have no means of determining, according to the standard of years, how long a period it takes to modify a species'.
13 Eicher, 1968, 22.
14 Jeletzky, 1956.

15 Hedberg, 1961a, 509.
16 *See*, for example, Simpson, 1963, 26; Shaw, 1964, 71.
17 *See* views quoted in Richmond, 1959, 675.
18 American Commission on Stratigraphic Nomenclature, 1961. Morpho-stratigraphical units are discussed below, chapter 7.
19 Flint, 1971, 372–4.
20 *ibid.*, 374.
21 For example, Morrison, 1968, 4.
22 Suggate, 1958–61.
23 *See* Ross, 1970.
24 Morrison, 1968, 75.
25 Emiliani, 1969.
26 Flint, 1965, xx.
27 Dimbleby, 1969, 167.
28 Renfrew, 1970, 207.

3 CHRONOMETRIC AGE

1 Fischer, 1969.
2 *See* examples in Flint, 1956.
3 For example, Higgs *et al.*, 1967.
4 Wheeler, 1954, 24.
5 Teichert, 1958, 108.
6 Callomon, 1966, quoted by Berry, 1968, 141.
7 For example, von Bubnoff, 1963, 117.
8 Kitts, 1966, 134.
9 Toulmin and Goodfield, 1967, 172 ff.
10 Reichenbach, 1958, 113 ff.
11 Carnap, 1966, 84–5; Borel, 1960, 74; cf. Whitrow, 1961, 44–6.
12 Faul, 1966, 78.
13 Shaw, 1964, 82–3.
14 Oakley, 1969, 35 n.
15 Holmes, 1963, xx–xxi.
16 Bell *et al.*, 1961.
17 Reichenbach, 1958, 128–9.
18 Kitts, 1966, 145.
19 Andrews, 1970, 20. De Geer, 1905 (cited by Lundqvist, 1965, 163), used 'equicesses' to denote geochronologically determined lines of ice recession.
20 Kobayashi, 1944, cited by Albritton, 1963, 294.
21 Schindewolf, 1957, 398; cf. Arkell, 1956, 461.
22 Evernden and Evernden, 1970, 83.
23 As does Kitts, 1966, 132.
24 Runcorn, 1970.
25 For a classic attempt *see* Zeuner, 1958.
26 This applies to some of the varves used by de Geer in his classic studies of Scandinavian glaciation (Hansen, 1965).
27 Faul, 1966, 9; Whitrow, 1961, 44–6; Holmes, 1947; von Bubnoff, 1963, 136.
28 *See* Wager, 1964.
29 For an example referring to the 'hard-water effect' *see* Shotton, 1972.
30 Vita-Finzi, 1970b.

31 Thurber, 1972, 1.
32 'Both radiocarbon and varve chronologies should be seen as variables and care should be taken in assessing the validity of one to justify or calibrate the other' (Johnson and Willis, 1970, 95).
33 Damon, Long and Grey, 1966, 1061.
34 J. C. Vogel *in* Olsson, 1970, 125–6.
35 Thurber, 1972, 1.
36 Broecker, 1965.
37 Dalrymple and Lanphere, 1969, 193–4.
38 For this and other techniques mentioned above *see* Brothwell and Higgs, 1969.
39 Andrews and Webber, 1969.
40 Mortimer, Clark and Schufle, 1971.
41 Waterbolk, 1971, 31.
42 Wasserburg, 1966, 440.
43 Faul, 1966, 40.
44 Wasserburg, 1966, 442.
45 cf. Cornford, 1922, 16.
46 Evernden and Evernden, 1970, 83.

4 SUBDIVIDING THE RECORD

1 Grabau, 1960, 1097.
2 Holmes, 1963, xviii.
3 Lyell, 1863, 6.
4 Lyell, 1833, 52.
5 Neustadt and Gudelis, 1965, 467.
6 For example, Sears, 1967.
7 Woodward, 1891, 419.
8 Ericson and Wollin, 1966. The exposition is marred by the authors' use of Recent to mean both postglacial times and, apparently, the entire Cainozoic.
9 Daly, 1934, 14 n. 13.
10 Lyell, 1833, 52.
11 Neustadt and Gudelis, 1965, 468.
12 Flint, 1971, 382–4.
13 West, 1968, 224–5.
14 Morrison *et al.*, 1957, 385.
15 Charlesworth, 1957, 596.
16 Morrison *et al.*, 1957, 385.
17 Morrison, 1969, 363.
18 Lyell, 1833, 52; 1863, 5.
19 Lyell, 1863, 6; cf. L. Agassiz, 1847, cited by Flint, 1971, 382.
20 Gillispie, 1959, 128.
21 Charlesworth, 1957, 1520.
22 Lundqvist, 1965, 171.
23 Reade, 1872.
24 Russell, 1958.
25 *See*, for example, Curray, 1965.
26 Van der Hammen, 1957, 250.
27 Broecker, Ewing and Heezen, 1960.
28 Hafsten, 1970; *see also* Mercer, 1972.

29   Curray, 1961, 1708.
30   Morrison, 1969, 366.
31   Flint, 1971, 383–4.
32   Suggate and West, 1967.
33   Hunt, 1965.
34   K. Keilhack, 1926, cited by Morrison *et al.*, 1957, 387.
35   Fairbridge and Newman, 1965, 414.
36   Anati *in* Emiliani, 1968, 30.
37   Coon *in* Emiliani, 1968, 33.
38   Oakley, 1964, 251.
39   Neustadt and Gudelis, 1965, 472; Schwerin *in* Emiliani, 1968, 41–2.
40   Martin and Wright, 1967; Kurtén, 1968.
41   Bourdier, 1962, 311; Gromov *et al.*, 1964.
42   Zeuner, 1959, 218.
43   Rankama, 1970, 410.
44   Morrison, 1969, 365.
45   Wickman, 1968, 318.

5   ANCIENT SHORELINES

1   Playfair, 1802, 457. cf. Axiom 53 of Scientology: 'A stable datum is necessary to the alignment of data'.
2   Chorley, 1963.
3   More precisely, 'The geoid is that equipotential surface of the earth the geometric height of which is most compatible with the surface determinable on the basis of the present local arithmetic means of sea-level heights in the open sea'. (Hela, 1969, 26–7.) For glacial effects *see* Fischer, 1959.
4   Vita-Finzi and Higgs, 1970.
5   Daly, 1934, 210.
6   Butzer, 1971a.
7   Nace, 1969.
8   Maclaren, 1842; Charlesworth, 1957, 1354–5; Donn, Farrand and Ewing, 1962.
9   McBurney and Hey, 1955; Vita-Finzi, 1969d, 606.
10   De Lamothe, 1899; Baulig, 1935.
11   Gill, 1967; Kuenen, 1955; Hill, 1962–3; Charlesworth, 1957, 1359; Schofield, 1967; Flint, 1971, 315–42. For a recent collection of sea-level studies *see Quaternaria*, 1971, **14**.
12   Zeuner, 1959, 290.
13   Zeuner, 1959, 302.
14   *See* Charlesworth, 1957, 1271.
15   Blanc, 1937, 623–4.
16   Zeuner, 1959, 302.
17   Balout, 1955, 39.
18   Flint, 1966.
19   Zeuner, 1959, 305; cf. Fairbridge, 1961, 121–2.
20   Friedman and Sanders, 1970.
21   Shackleton, 1969.
22   Emiliani and Rona, 1969; Broecker and Ku, 1969.
23   Chorley, 1963, 964.
24   Newell and Bloom, 1970; cf. Zeuner, 1959.

25 Gigout, 1962, 213.
26 Playfair, 1802, 446.
27 Mörner, 1969.
28 Walcott, 1972.
29 Munk and MacDonald, 1960, 248; Hipkin, 1970.
30 Charlesworth, 1957, 1494; this passage is also cited by Newell and Bloom, 1970, 1882.
31 Curray, 1961; van Andel *et al.*, 1967.
32 Curray, 1965, 725; Jelgersma, 1966.
33 Schnable and Goodell, 1968.
34 Bloom, 1971, 375.
35 Hela, 1969, 41. *See also* Fairbridge, 1963, figure 4 for disparities in the record of the Indian, Pacific and Atlantic for 1860–1960.
36 Fairbridge, 1966, 481–2.
37 Munk and MacDonald, 1960, 234.
38 Daly, 1934, 157–64; Fairbridge, 1961, 161–73 puts 'fresh' in inverted commas.
39 Shepard and Curray, 1967.
40 Bloom, 1964; Wiggers, 1954, 179.
41 For the problems raised by recrystallisation of shell carbonates *see* Chappell and Polach, 1972.
42 Coleman and Smith, 1964; Scholl and Stuiver, 1967.
43 The point is made by Jelgersma, 1966, 58, about the curve in Fairbridge, 1961. This is based on 'Stable and delta areas'—strange bedfellows.
44 For example, Tipasa (− 3990 ± 200) *in* Fairbridge, 1961. Florida (− 3800 ± 200) is similarly adjusted by invoking compaction.
45 Fujii and Fuji, 1967, accept a + 6 m level between 7000 and 3000 years ago with no more qualification than 'we cannot neglect to consider local movements'.
46 Battistini, 1963; Guilcher, 1969, 73.
47 Daly, 1934, 164–5, 293.
48 Godwin, Suggate and Willis, 1958.
49 Jelgersma, 1966, 60.
50 Fairbridge, 1961, 163.
51 Anderson and Brückner, 1965, 258.
52 Bloom, 1970.
53 Curray, Shepard and Veeh, 1970. *See also* Shepard, 1970; Newell and Bloom, 1970.
54 Newman and Rusnak, 1965, cited approvingly by Kraft, 1971, 2154.
55 Vita-Finzi, 1970a.
56 Pearson, 1901, 167.
57 Gill, 1966, 407.
58 Wiener, 1965, 44; Dobzhansky, 1970, 58.
59 Gilbert, 1890; Crittenden, 1963.
60 Morrison, 1965.
61 Andersen, 1965, 127.
62 Andrews, 1970, 98.
63 West, 1968, 145–7.
64 Lundqvist, 1965, 181.
65 Zeuner, 1958, 55.
66 Andrews, 1970.

67 Churchill, 1965.
68 Smalley, 1967.
69 Flemming, 1969.
70 Flemming, 1968.

6 CORES AND BOREHOLES

1 West, 1968, 129–31; Higgs and Webley, 1971.
2 Davis and Deevey, 1964.
3 Dimbleby, 1969.
4 The absence of a tree cover around a prehistoric site in Greece was explained to the writer as the product of Neolithic deforestation. Although there was no evidence for such a cover in the past, the argument was that, elsewhere in Europe, Neolithic man had habitually engaged in clearing.
5 Deevey, 1965, 646.
6 Raikes, 1967, 96–100.
7 Deevey, 1965, 648.
8 Keen, 1968, 53.
9 Phleger, 1960, 245.
10 Olausson, 1965; Broecker, 1965.
11 Ericson et al., 1961, 282, 244.
12 Ericson, 1963.
13 Emiliani, 1969.
14 Shackleton, 1967.
15 See Shackleton and Turner, 1967.
16 Langway, 1970; Dansgaard et al., 1971.
17 Koczy, 1963.
18 Opdyke et al., 1966.
19 Broecker et al., 1958.
20 Rankama, 1965.
21 Frerichs, 1968.
22 Sears, 1967.
23 Pennington, 1969, 64–5.
24 Heusser, 1966.
25 F. Oldfield in Sawyer, 1966, 141.
26 Damuth and Fairbridge, 1970; Groot et al., 1967.
27 Vita-Finzi, 1973.
28 Rossignol, 1969; Vita-Finzi and Dimbleby, 1971.
29 Godwin, 1966, 12; McBurney, 1967.
30 M. I. Finley, Review of J. W. Mavor, Jr, *Voyage to Atlantis* (1969), *New York Review of Books*, 1969, **12**, 38–40.
31 Ninkovich and Heezen, 1965.
32 Wyllie, 1971, 353.
33 Frakes and Kemp, 1972.

7 RIVER DEPOSITS

1 Forbes, 1963.
2 Chorley et al., 1964, 190.
3 Lyell, 1849, cited by Chorley et al., 1964, 268; Vita-Finzi, 1969a.

4 Frye and Willman, 1962.
5 Richmond, 1962.
6 American Commission on Stratigraphic Nomenclature, 1961.
7 Leopold *et al.*, 1964, 442.
8 *ibid.*, 465–7.
9 Vita-Finzi, 1964; Farrand, 1971.
10 Oakley, 1964, 93.
11 Schumm, 1965.
12 Meier, 1965, 801.
13 Chorley, 1963.
14 Leopold *et al.*, 1964, 260–1, 454–6.
15 Anderson, 1936; Vita-Finzi, 1967.
16 Vita-Finzi, 1968.
17 Miller and Scott, 1961.
18 Flint, 1971, 525.
19 Williams, 1970.
20 Fairbridge, 1962; cf. Sandford and Arkell, 1933.
21 Berry and Whiteman, 1968.
22 Haynes, 1968.
23 Judson, 1963; Vita-Finzi, 1969b,d,e.
24 cf. Wallén, 1956, for a Mexican parallel.
25 Cooke, 1972.
26 Leopold and Miller, 1954; Ruhe and Daniels, 1965; Miller and Wendorf, 1958.
27 Leopold and Miller, 1954, 72–3.
28 The fact that the Lantien mandible from Shensi Province lay under 80 m of clay is taken to show that it is 'ancient' (*New Scientist*, 1964, **22**, 400–1).
29 Hudson, 1964, 41.
30 For example, Edelman, 1956; Butzer, 1970.
31 Larrabee, 1962.
32 Vita-Finzi, 1966.
33 Miller and Wendorf, 1958, 183–4.
34 Vita-Finzi and Smalley, 1970; Vita-Finzi, 1971.
35 Oakley, 1964, 7–8.
36 Coles and Higgs, 1969, 417.
37 Butzer, 1971a, 229–31, and *passim*.
38 Higgs and Vita-Finzi, 1966; Higgs *et al.*, 1967; Harris and Vita-Finzi, 1968.
39 Vita-Finzi, 1970b. For the Oaxaca area *see* Flannery *et al.*, 1967.
40 Byers, 1967.
41 Raikes, 1967.
42 Lyell, 1872, **1**, 434.
43 Coque, 1962; Bishop and Clark, 1967, 357.
44 Wellman, 1955.
45 Zak and Freund, 1966.
46 Williams, 1970, 58.

8 CAVE INFILLS

1 Laville, 1964; Bordes, 1969.
2 Brain, 1967. For a recent reassessment *see* Butzer, 1971b.

3   Garrod and Bate, 1937.
4   Woldstedt, 1967, 285.
5   Schmid, 1969, 160.
6   Frank, 1971; Butzer, 1971a, 208–14.
7   Zeuner, 1959; Butzer, 1957.
8   McBurney, 1967.
9   Oakley, 1964, 32, 109.
10   Butzer, 1971a, 44; Ericson and Wollin, 1966, 216.
11   Zeuner, 1959; Choubert, 1961, 14.
12   Bishop, 1967.
13   Ewer, 1967.
14   Cooke, 1967.
15   Kurtén, 1968, 15–17.
16   Morrison, 1968.
17   *See*, for example, the tables in Flint, 1971, 748–50.
18   Bordes, 1968, 17–18.
19   Clark, 1969, 46, 66.
20   Bishop and Clark, 1967, *passim*.
21   McBurney, 1967, 67.
22   Coles and Higgs, 1969, 221.
23   Ewer, 1967, 119.
24   Kurtén, 1968, 244.
25   Martin and Wright, 1967.
26   Higgs, 1961, 144.
27   Sackett, 1968, 67.
28   Coles and Higgs, 1969, 271–2; Bordes, 1969, 47. Sackett, 1968, 79 n.
     10, states that the Aurignacian V and Proto-Magdalenian phases are
     presently known from a single and two occupations respectively.
29   Tournal, 1833.
30   Poulson and White, 1969.
31   Ellwood, 1971.
32   Kurtén, 1964.
33   Solecki and Leroi-Gourhan, 1961, 734.
34   Kowalski, 1971.
35   Tchernov, 1968.
36   Vita-Finzi and Higgs, 1970.
37   Higgs *et al.*, 1967.
38   M. Harris *in* Binford and Binford, 1968, 360.

9   CASES AND LAWS

1   The (anonymous) comments refer to an early draft of this book.
2   Wheeler, 1954, 24–5, 39, 215.
3   Vita-Finzi, 1969c.
4   Wright, 1971, 443.
5   Mercer, 1969; Manley, 1971.
6   M. Gluckman, review of L. Tiger and R. Fox, *The Imperial Animal*
     (1972), *New York Review of Books*, 1972, **19**, 41.
7   Flint, 1959; Bishop, 1968.
8   cf. Koestler, 1967, 391–5.
9   Page, 1972.
10   Medawar, 1969, 128.

11 Medawar, 1969, 84, 91. I cannot resist quoting my anonymous reviewer again: 'Palaeontology—and particularly de Chardin's very limited or specialised kind—and stratigraphy *are* humble and intellectually unexacting and nothing written by [the author] or anyone else can change this for a long time'.

12 *See*, for example, the essays in Teilhard de Chardin, 1965.

13 Sylvester-Bradley, 1967, 52.

14 Wilson, 1959, quoted by Morrison, 1968, 11.

15 Kuhn, 1970. Connoisseurs of the fashionable will be pleased by this reference. Those who like having it both ways will appreciate this note.

16 Lyell, 1830, 1, 164–5.

17 Gould, 1965.

18 cf. Medawar, 1969, 147.

19 Watson, 1966.

20 Bucher, 1933, 148n.

21 Chorley, 1966, 278.

22 Schumm and Lichty, 1965.

23 Thompson, 1961, 47–8.

24 Scheidegger, 1963, 160–1.

25 Russell, 1953, 193–4; *see also* Whitrow, 1961, 174.

26 Collingwood, 1961, 212–5.

27 Johnson, 1955, 142.

28 Cited by Wheeler, 1954, 205.

Bibliography

Adams, F. D. (1954). *The Birth and Development of the Geological Sciences,* Dover, New York.

Albritton, C. C. Jr (ed.) (1963). *The Fabric of Geology,* Addison-Wesley, Reading, Mass.

American Commission on Stratigraphic Nomenclature (1961). Code of Stratigraphic Nomenclature. *Bull. Am. Ass. Petrol. Geol.,* **45**, 645–60.

Andel, Tj. H. van, Heath, G. R., Moore, T. C., and McGeary, D. F. R. (1967). Late Quaternary history, climate, and oceanography of the Timor Sea, northwestern Australia. *Am. J. Sci.,* **265**, 737–58.

Andersen, B. G. (1965). The Quaternary of Norway. *In* Rankama, 1965, 91–138.

Anderson, M. M., and Brückner, W. D. (1965). The raised shore-lines of Ghana, West Africa. *Rep. VI Int. Cong. Quat. (INQUA) Warsaw, 1961,* Lódz, 253–68.

Anderson, R. van V. (1936). Geology in the coastal Atlas of western Algeria. *Mem. geol. Soc. Am.,* **4**.

Andrews, J. T. (1970). A geomorphological study of postglacial uplift with particular reference to Arctic Canada. *Spec. Publ. Inst. Brit. geog., Lond.,* **2**.

Andrews, J. T., and Webber, P. J. (1969). A lichenometrical study of the northwestern margin of Barnes Ice Cap: a geomorphological technique. *In Geomorphology* (eds. J. G. Nelson and M. J. Chambers), Methuen, London, pp. 65–88.

Arkell, W. J. (1956). Comments on stratigraphic procedure and terminology. *Am. J. Sci.,* **254**, 457–67.

Balout, L. (1955). *Préhistoire de l'Afrique du Nord.* Arts et Métiers Graphiques, Paris.

Battistini, R. (1963). L'âge absolu de l'encoche de corrosion marine fland-rienne de 1–1, 3m de la baie des Galions (extrême-sud de Madagascar). *C. R. Soc. géol., Fr.,* 16–17.

Baulig, H. (1935). The changing sea level. *Publ. Inst. Brit. geog., Lond.,* **3**.

Bell, W. C., Kay, M., Grover, E. M., Wheeler, H. E., and Wilson, J. A. (1961). Stratigraphic Commission. Note 25—Geochronologic and chrono-stratigraphic units. *Bull. Am. Ass. Petrol. Geol.,* **45**, 666–73.

Berry, L., and Whiteman, A. J. (1968). The Nile in the Sudan. *Geogrl. J.,* **134**, 1–37.

Berry, W. B. N. (1968). *Growth of a Prehisoric Time Scale,* W. H. Freeman and Co., San Francisco and London.

Binford, S. R., and Binford, L. R. (eds.) (1968). *New Perspectives in Archeology,* Aldine, Chicago.

Bishop, W. W. (1967). Annotated lexicon of Quaternary stratigraphical nomenclature in East Africa. *In* Bishop and Clark, 1967, 375–95.

Bishop, W. W. (1968). Means of correlation of Quaternary successions in East Africa. *In* Morrison and Wright, 1968, 161–72.

Bishop, W. W., and Clark, J. D. (eds.) (1967). *Background to Evolution in Africa*, University of Chicago Press, Chicago and London.

Blanc, A. C. (1937). Low levels of the Mediterranean Sea during the Pleistocene glaciation. *Q. Jl geol. Soc. Lond.*, **93**, 621–51.

Bloom, A. L. (1964). Peat accumulation and compaction in a Connecticut coastal marsh. *J. sedim. Petrol.*, **34**, 599–603.

Bloom, A. L. (1970). Paludal stratigraphy of Truk, Ponape, and Kusaie, Eastern Caroline Islands. *Bull. geol. Soc. Am.*, **81**, 1895–904.

Bloom, A. L. (1971). Glacial-eustatic and isostatic controls of sea level since the Last Glaciation. *In* Turekian, 1971, 355–79.

Bordes, F. (1968). *The Old Stone Age*, Weidenfeld and Nicolson, London.

Bordes, F. (1969). Landes Périgord. Livret Guide de l'Excursion A5, *INQUA Congress 8, Paris*. Biscaye Frères, Bordeaux.

Borel, E. (1960). *Space and time*, Dover, New York.

Bourdier, F. (1962). *Le Bassin du Rhône au Quaternaire* (2 vols.), Louis-Jean, Gap.

Brain, C. K. (1967). Procedures and some results in the study of Quaternary cave fillings. *In* Bishop and Clarke, 1967, 285–301.

Broecker, W. S. (1965). Isotope geochemistry and the Pleistocene climatic record. *In* Wright and Frey, 1965, 737–53.

Broecker, W. S., Ewing, M., and Heezen, B. C. (1960). Evidence for an abrupt change in climate close to 11 000 years ago. *Am. J. Sci.*, **258**, 429–48.

Broecker, W. S., and Ku, T. L. (1969). Caribbean cores P6304–8 and P6304–9: new analysis of absolute chronology. *Science, N.Y.*, **166**, 404–6.

Broecker, W. S., Turekian, K. K., and Heezen, B. C. (1958). The relation of deep sea sedimentation rates to variations in climate. *Am. J. Sci.*, **256**, 503–17.

Brothwell, D., and Higgs, E. (eds.) (1969). *Science in Archaeology*, Thames and Hudson, London.

Brunet, J. (1967). Geologic studies. *In* Byers, 1967, 66–90.

Bryson, R. A., Wendland, W. M., Ives, J. D., and Andrews, J. T. (1969). Radiocarbon isochrones on the disintegration of the Laurentide ice sheet. *Arctic and Alpine Res.*, **1**, 1–14.

Bubnoff, S. von (1963). *Fundamentals of Geology* (translated and edited by W. T. Harry), Oliver and Boyd, Edinburgh and London.

Bucher, W. H. (1933). *The Deformation of the Earth's Crust*, Princeton University Press, Princeton.

Butzer, K. W. (1957). Mediterranean pluvials and the general circulation of the atmosphere. *Geogr. Annlr*, **39**, 48–53.

Butzer, K. W. (1970). Contemporary depositional environments of the Omo delta. *Nature, Lond.*, **226**, 425–30.

Butzer, K. W. (1971a). *Environment and Archeology* (2nd edition), Methuen, London.

Butzer, K. W. (1971b). Another look at the Australopithecine cave breccias of the Transvaal. *Am. Anthrop.*, **73**, 1197–201.

Byers, D. S. (ed.) (1967). *The Prehistory of the Tehuacan Valley. I: Environment and Subsistence*, University of Texas Press, Austin.

Carnap, R. (1966). *Philosophical Foundations of Physics*, Basic Books, New York.

Chappell, J., and Polach, H. A. (1972). Some effects of partial recrystallisation on ^{14}C dating late Pleistocene corals and molluscs. *Quaternary Research*, **2**, 244–52.

Charlesworth, J. K. (1957). *The Quaternary Era* (2 vols.), Arnold, London.

Chorley, R. J. (1963). Diastrophic background to twentieth-century geomorphological thought. *Bull. geol. Soc. Am.*, **74**, 953–70.

Chorley, R. J. (1966). The application of statistical methods to geomorphology. *In Essays in Geomorphology* (ed. G. H. Dury), Heinemann, London, pp. 275–387.

Chorley, R. J., Dunn, A. J., and Beckinsale, R. P. (1964). *The History of the Study of Landforms: I*, Methuen, London.

Choubert, G. (1961). Quaternaire du Maroc. *Biul. peryglac.* **10**, 9–29.

Churchill, D. M. (1965). The displacement of deposits formed at sea-level 6500 years ago in Southern Britain. *Quaternaria*, **7**, 239–49.

Clark, G. (1969). *World Prehistory* (2nd edition), Cambridge University Press.

Coleman, J.M., and Smith, W. G. (1964). Late Recent rise of sea level. *Bull, geol. Soc. Am.*, **75**, 833–40.

Coles, J. M., and Higgs, E. S. (1969). *The Archaeology of Early Man*, Faber and Faber, London.

Collingwood, R. G. (1961). *The Idea of History* (1st edition 1946), Oxford University Press.

Cooke, H. B. S. (1967). The Pleistocene sequence in South Africa and problems of correlation. *In* Bishop and Clark, 1967, 175–84.

Cooke, R. U. (1972). A model of arroyo development in historical times. *Int. Geog. Cong., Colloque de Ouargla 1971*, **2**, 1–8.

Coque, R. (1962). *La Tunisie Présaharienne*, Colin, Paris.

Cornford, F. M. (1922). *Microcosmographia Academica* (2nd edition), Bowes and Bowes, Cambridge.

Crittenden, M. D., Jr. (1963). New data on the isostatic deformation of Lake Bonneville. *Prof. Pap. U.S. geol. Surv. 454–E*.

Cserna, Z. de (1972). Review of 'Stratigraphie und Stratotypus'; by Otto H. Schindewolf. *Am. J. Sci.*, **272**, 189–94.

Curray, J. R. (1961). Late Quaternary sea level: a discussion. *Bull. geol. Soc. Am.*, **72**, 1707–12.

Curray, J. R. (1965). Late Quaternary history, continental shelves of the United States. *In* Wright and Frey, 1965, 723–35.

Curray, J. R., Shepard, F. P., and Veeh, H. H. (1970). Late Quaternary sea-level studies in Micronesia: CARMARSEL Expedition. *Bull. geol. Soc. Am.*, **81**, 1865–80.

Dalrymple, G. B., and Lanphere, M. A. (1969). *Potassium–Argon dating*, Freeman, San Francisco.

Daly, R. A. (1934). *The Changing World of the Ice Age*, Yale University Press, New Haven.

Damon, P. E., Long, A., and Grey, D. C. (1966). Fluctuation of atmospheric ^{14}C during the last six millennia. *J. geophys. Res.*, **71**, 1055–63.

Damuth, J. E., and Fairbridge, R. W. (1970). Equatorial Atlantic deep-sea arkosic sands and ice-age aridity in Tropical South America. *Bull. geol. Soc. Am.*, **81**, 189–206.

Dansgaard, W., Johnsen, S. J., Clausen, H. B., and Langway, C. C., Jr. (1971). Climatic record revealed by the Camp Century ice core. *In* Turekian, 1971, 37–56.

Davis, M. B., and Deevey, E. S., Jr. (1964). Pollen accumulation rates: estimates from late-glacial sediment of Rogers Lake. *Science, N.Y.*, **145**, 1293–5.

Deevey, E. S., Jr. (1965). Pleistocene nonmarine environments. *In* Wright and Frey, 1965, 643–52.

Dimbleby, G. W. (1969). Pollen analysis. *In* Brothwell and Higgs, 1969, 167–77.

Dobzhansky, T. (1970). *Mankind Evolving* (1st edition 1962), Bantam Books, Toronto.

Donn, W. L., Farrand, W. F., and Ewing, M. (1962). Pleistocene ice volumes and sea-level lowering. *J. Geol.*, **70**, 206–14.

Donovon, D. T. (1966). *Stratigraphy*, Thomas Murby, London.

Edelman, C. H. (1956). Sedimentology of the Rhine and Meuse delta as an example of the sedimentology of the Carboniferous. *Verh. Geol. Mijnb. Gen. Geol. ser. XVI*, 64–75.

Eicher, D. L. (1968). *Geologic Time*, Prentice-Hall, Englewood Cliffs.

Ellwood, B. B. (1971). An archeomagnetic measurement of the age and sedimentation rate of Climax Cave sediments, southwest Georgia. *Am. J. Sci.*, **271**, 304–10.

Emiliani, C. (1955). Pleistocene temperatures. *J. Geol.*, **63**, 538–78.

Emiliani, C. (1968). The Pleistocene Epoch and the evolution of Man. *Curr. Anthrop.* **9**, 27–47.

Emiliani, C. (1969). The significance of deep-sea cores. *In* Brothwell and Higgs, 1969, 109–117.

Emiliani, C., and Rona, E. (1969). Caribbean cores P6304-8 and P6304-9: new analysis of absolute chronology. A reply. *Science, N.Y.*, **166**, 1551–2.

Ericson, D. B. (1963). Cross-correlation of deep-sea sediment cores and determination of relative rates of sedimentation by micropaleontological techniques. *In* Hill, 1962–3, **3**, 832–42.

Ericson, D. B., Ewing, M., Wollin, G., and Heezen, B. C. (1961). Atlantic deep-sea sediment cores. *Bull. geol. Soc. Am.*, **72**, 193–286.

Ericson D. B., and Wollin, G. (1966). *The Deep and the Past*, Jonathan Cape, London.

Evernden, J. F., and Evernden, R. K. S. (1970). The Cenozoic time scale. *Spec. Pap. geol. Soc. Am.*, **124**, 71–90.

Ewer, R. F. (1967). The fossil hyaenids of Africa—a reappraisal. *In* Bishop and Clark, 1967, 109–123.

Fairbridge, R. W. (1961). Eustatic changes in sea level. *In Physics and Chemistry of the Earth*, vol. 4 (ed. L. H. Ahrens), Pergamon Press, New York, pp. 99–185.

Fairbridge, R. W. (1962). New radiocarbon dates of Nile sediments. *Nature, Lond.*, **196**, 108–10.

Fairbridge, R. W. (1963). Mean sea level related to solar radiation during the last 20 000 years. *In Changes of Climate, UNESCO, Paris*, pp. 229–42.

Fairbridge, R. W. (ed.) (1966). *The Encyclopedia of Oceanography*, Reinhold, New York.

Fairbridge, R. W., and Newman, W. S. (1965). Sea level and the Holocene boundary in the eastern United States. *Rep. VI. Int. Cong. Quat. (INQUA)* Warsaw, *Łódz*, **1**, 397–418.

Farrand, W. R. (1971). Late Quaternary palaeoclimates of the eastern Mediterranean area. *In* Turekian, 1971, 529–64.

Faul, H. (1966). *Ages of Rocks, Planets, and Stars*, McGraw-Hill, New York.

Fischer, A. G. (1969). Geological time-distance rates: the Bubnoff Unit. *Bull. geol. Soc. Am.*, **80**, 549–51.

Fischer, I. (1959). The impact of the Ice Age on the present form of the geoid. *J. geophys. Res.*, **64**, 85–87.

Flannery, K. V., Kirkby, A. V. T., Kirkby, M. J., and Williams, A. W., Jr. (1967). Farming systems and political growth in ancient Oaxaca. *Science, N.Y.*, **158**, 445–53.

Flemming, N. C. (1968). Holocene earth movements and eustatic sea level change in the Peloponnese. *Nature, Lond.*, **217**, 1031–2.

Flemming, N. C. (1969). Archaeological evidence for eustatic change of sea level and earth movements in the western Mediterranean during the last 2000 years. *Spec. Pap. geol. Soc. Am.*, **109**.

Flint, R. F. (1956). New radiocarbon dates and late-Pleistocene stratigraphy. *Am. J. Sci.*, **254**, 265–87.

Flint, R. F. (1959). On the basis of Pleistocene correlation in East Africa. *Geol. Mag.*, **96**, 265–84.

Flint, R. F. (1965). Introduction. *In* Rankama, 1965, xi–xxii.

Flint, R. F. (1966). Comparison of interglacial marine stratigraphy in Virginia, Alaska, and Mediterranean areas. *Am. J. Sci.*, **264**, 673–84.

Flint, R. F. (1971). *Glacial and Quaternary Geology*, Wiley, New York.

Forbes, R. J. (1963). *Studies in Ancient Technology VII*, Brill, Leiden.

Frakes, L. A., and Kemp, E. M. (1972). Generation of sedimentary facies on a spreading ocean ridge. *Nature, Lond.*, **236**, 114–17.

Frank, R. (1971). Cave sediments as palaeoenvironmental indicators, and the sedimentary sequence in Koonalda Cave. *In Aboriginal Man and Environment in Australia* (eds. D. J. Mulvaney and J. Golson), Australian National University Press, Canberra, pp. 94–104.

Frerichs, W. E. (1968). Pleistocene–Recent boundary and Wisconsin glacial biostratigraphy in the northern Indian Ocean. *Science, N.Y.*, **159**, 1456–8.

Friedman, G. M., and Sanders, J. E. (1970). Coincidence of high sea level with cold climate and low sea level with warm climate: evidence from carbonate rocks. *Bull. geol. Soc. Am.*, **81**, 2457–8.

Frye, J. C., and Willman, H. B. (1962). Morphostratigraphic units in Pleistocene stratigraphy. *Bull Am. Ass. Petrol. Geol.*, **46**, 112–3.

Fujii, S., and Fuji, N. (1967). Postglacial sea level in the Japanese islands. *Jour. Geosci., Osaka City Univ.*, **10**, 43–51.

Garrod, D. A. E., and Bate, D. M. A. (1937). *The Stone Age of Mount Carmel: I*, Clarendon Press, Oxford.

Gigout, M. (1962). Sur le Tyrrhénian de la Méditérranée occidentale. *Quaternaria*, **6**, 209–28.

Gilbert, G. K. (1890). Lake Bonneville. *Mon. U.S. geol. Surv.*, **1**.

Gill, E. D. (ed.) (1966). Australasian research in Quaternary shorelines. *Aust. J. Sci.*, **28**, 407–11.

Gill, E. D. (1967). Criteria for the description of Quaternary shorelines. *Quaternaria*, **9**, 237–43.

Gillispie, C. C. (1959). *Genesis and Geology*, Harper and Row, New York.

Godwin, H. (1966). Introductory address. *In* Sawyer, 1966, 3–14.

Godwin, H., Suggate, R. P., and Willis, E. H. (1958). Radiocarbon dating of the eustatic rise in ocean level. *Nature, Lond.*, **181**, 1518.

Gould, S. J. 1965). Is uniformitarianism necessary? *Am. J. Sci.*, **263**, 223–8.

Grabau, A. (1960). *Principles of Stratigraphy* (revised edition, first published in 1924), 2 vols., Dover, New York.

Gromov, V. I., Krasnov, I. I., Nikiforova, K. V., and Shancer, E. V. (1964). State of the problem of the lower boundary and the stratigraphic sub-division of the Anthropogene (Quaternary) System. *Rep. VI Int. Cong. Quat. (INQUA), Warsaw 1961, Lódz*, **2**, 95–104.

Groot, J. J., Groot, C. R., Ewing, M., Burckle, L., and Conolly, J. R. (1967). Spores, pollen, diatoms and provenance of Argentine Basin sediments. *In* Sears, 1967, 179–217.

Guilcher, A. (1969). Pleistocene and Holocene sea level changes. *Earth-Sci. Rev.*, **5**, 69–97.

Hafsten, U. (1970). A sub-division of the late Pleistocene period on a syn-chronous basis, intended for global and universal usage. *Palaeogeography, Palaeoclimatol., Palaeoecol.*, **7**, 279–96.

Hammen, T. van der (1967). The stratigraphy of the Late-glacial. *Geologie Mijnb.*, **19**, 250–4.

Hansen, S. (1965). The Quaternary of Denmark. *In* Rankama, 1965, 1–90.

Harland, W. B., Smith, A. G., and Wilcock, B. (eds.) (1964). The Phanero-zoic time-scale. *Q. Jl geol. Soc. Lond.*, **120**s.

Harris, D. R., and Vita-Finzi, C. (1968). Kokkinopilos—a Greek badland. *Geogrl. J.*, **134**, 537–46.

Haynes, C. V., Jr. (1968). Geochronology of late-Quaternary alluvium. *In* Morrison and Wright 1968, 591–631.

Hedberg, H. D. (1961a). The stratigraphic panorama. *Bull. geol. Soc. Am.*, **72**, 499–518.

Hedberg, H. D. (ed.) (1961b). Stratigraphic Classification and Terminology. *Rep. 21 Int. Geol. Cong. (Norden) XXV*, Copenhagen.

Hela, I. (1969). Mean sea level. *In Oceanography*. Annals of the Inter-national Geophysical Year, Vol. XLVI (eds. A. L. Gordon and F. W. G. Baker). Pergamon, Oxford, 25–45.

Heusser, C. J. (1966). Polar hemispheric correlation: palynological evidence from Chile and the Pacific north-west of America. *In* Sawyer, 1966, 124–41.

Higgs, E. S. (1961). Some Pleistocene faunas of the Mediterranean coastal areas. *Proc. prehist. Soc.*, **27**, 144–154.

Higgs, E. S., and Vita-Finzi, C. (1966). The climate, environment and industries of Stone Age Greece: Part II. *Proc. prehist. Soc.*, **32**, 1–29.

Higgs, E. S., Vita-Finzi, C., Harris, D. R., and Fagg, A. E. (1967). The climate, environment and industries of Stone Age Greece: Part III. *Proc. prehist. Soc.*, **33**, 1–29.

Higgs, E. S., and Webley, D. (1971). Further information concerning the environment of Palaeolithic man in Epirus. *Proc. prehist. Soc.*, **37**, 367–80.

Hill, M. N. (ed.) (1962–3). *The Sea* (3 vols.), Interscience, New York.

Hipkin, R. G. (1970). A review of the theories of the earth's rotation. *In* Runcorn, 1970, 53–9.

Holmes, A. (1947). The construction of a geological time-scale. *Trans. geol. Soc., Glasgow*, **21**, 117–52.

Holmes, A. (1963). Introduction. *In* Rankama, 1965, xi–xxiv.

Huang, T. C., Stanley, D. J., and Stuckenrath, R. (1972). Sedimentological evidence for current reversal at the Strait of Gibraltar. *Marine Technol. J.*, **6**, 25–33.

Hudson, J. D. (1964). Sedimentation rates in relation to the Phanerozoic time-scale. *In* Harland *et al.*, 1964, 37–42.

Hunt, C. B. (1965). Quaternary geology reviewed. *Science, N.Y.*, **150**, 47–50.

Huxley, T. H. (1862). Anniversary address. *Q. Jl geol. Soc. Lond.*, **18**, xl–liv.

Interdepartmental Stratigraphic Committee of the USSR (1966). Stratigraphic classification, terminology and nomenclature. (Published in USSR in 1965). *Int. Geol. Rev.*, **10**, 1–36 (English translation).

Jeletzky, J. A. (1956). Paleontology, basis of practical geochronology. *Bull. Am. Ass. Petrol. Geol.*, **40**, 679–706.

Jelgersma, S. (1966). Sea-level changes during the last 10 000 years. *In* Sawyer, 1966, 54–71.

Johnson, F. (1955). Reflections upon the significance of radiocarbon dates. *In Radiocarbon Dating* (2nd edition) by W. F. Libby, University of Chicago Press, Chicago, pp. 141–61.

Johnson, F., and Willis, E. H. (1970). Reconciliation of radiocarbon and sidereal years in Meso-American chronology. *In* Olsson, 1970, 93–104.

Judson, S. (1963). Geology of the Hodges Sites, Quay County, New Mexico. *Bull. Bur. Am. Ethnol.*, **154**, 285–302.

Keen, M. J. (1968). *An Introduction to Marine Geology*, Pergamon, Oxford.

Kitts, D. B. (1966). Geologic time. *J. Geol.*, **74**, 127–46.

Koczy, F. F. (1963). Age determination in sediments by natural radioactivity. *In* Hill, 1962–3, vol. 3, 816–31.

Koestler, A. (1967). *The Ghost in the Machine*, Pan Books, Aylesbury.

Kowalski, K. (1971). The biostratigraphy and paleoecology of late Cenozoic mammals of Europe and Asia. *In* Turekian, 1971, 465–77.

Kraft, J. C. (1971). Sedimentary facies patterns and geologic history of a Holocene marine transgression. *Bull. geol. Soc. Am.*, **82**, 2131–58.

Krumbein, W. C., and Sloss, L. L. (1963). *Stratigraphy and Sedimentation* (2nd edition), Freeman, San Francisco.

Kuenen, P. H. (1955). Sea level and crustal warping. *In* Crust of the Earth (ed. A. Poldervaart), *Spec. Pap. geol. Soc. Am.*, **62**, 193–204.

Kuhn, T. S. (1970). *The Structure of Scientific Revolutions* (2nd edition), University of Chicago Press, Chicago.

Kurtén, B. (1964). Population structure in paleoecology. *In Approaches to Paleoecology* (eds. J. Imbrie and N. Newell), Wiley, New York, pp. 91–106.

Kurtén, B. (1968). *Pleistocene Mammals of Europe*, Weidenfeld and Nicolson, London.

Lamothe, L. J. B. de (1899). Note sur les anciennes plages et terrasses du bassin de l'Isser (Département d'Alger) et de quelques autres bassins de la côte algérienne. *Bull. Soc. géol. Fr.*, **27**, 257–303.

Langway, C. C., Jr. (1970). Stratigraphic analysis of a deep ice core from Greenland. *Spec. Pap. geol. Soc. Am.*, **125**.

Larrabee, E. McM. (1962). Ephemeral water action preserved in closely dated deposit, *J. sedim. Petrol.*, **32**, 608–9.

Laville, H. (1964). Recherches sédimentologiques sur la paléoclimatologie du Wurmien récent en Périgord. *L'Anthropologie, Paris*, **68**, 1–48, 219–52.

Leopold, L. B., and Miller, J. P. (1954). A Postglacial chronology for some alluvial valleys in Wyoming. *Prof. Pap. U.S. geol. Surv., 1261*.

Leopold, L. B., Wolman, M. G., and Miller, J. P. (1964). *Fluvial Processes in Geomorphology*, W. H. Freeman, San Francisco.

Lundqvist, J. (1965). The Quaternary of Sweden. *In* Rankama, 1965, 139–98.

Lyell, C. (1830–3). *Principles of Geology* (vol. 1 : 1830; vol. 2 : 1832; vol. 3 : 1833), Murray, London.

Lyell, C. (1863). *The Geological Evidences of the Antiquity of Man*, Murray, London.

Lyell, C. (1872). *Principles of Geology* (11th edition), 2 vols., Murray, London.

McBurney, C. B. M. (1967). *The Haua Fteah (Cyrenaica)*, Cambridge University Press.

McBurney, C. B. M., and Hey, R. W. (1955). *Prehistory and Pleistocene Geology in Cyrenaican Libya*, Cambridge University Press.

Maclaren, C. (1842). The glacial theory of Professor Agassiz. *Am. J. Sci.*, **42**, 346–65.

Manley, G. (1971). Interpreting the meteorology of the Late and Post-glacial. *Palaeogeography, Palaeoclimatol., Palaeoecol.*, **10**, 163–75.

Martin, P. S., and Wright, H. E., Jr. (eds.) (1967). *Pleistocene Extinctions*, Yale University Press, New Haven.

Medawar, P. B. (1969). *The Art of the Soluble*, Penguin Books, Harmondsworth.

Meier, M. F. (1965). Glaciers and Climate. *In* Wright and Frey, 1965. 795–805.

Mercer, J. H. (1969). The Allerød oscillation: a European climatic anomaly? *Arctic and Alpine Research*, **1**, 227–34.

Mercer, J. H. (1972). The lower boundary of the Holocene. *Quaternary Research*, **2**, 15–24.

Miller, J. P., and Wendorf, F. (1958). Alluvial chronology of the Tesuque valley, New Mexico. *J. Geol.*, **66**, 177–94.

Miller, R. D., and Scott, G. R. (1961). Late Wisconsin age of terrace alluvium along the North Loup River, Central Nebraska: a revision. *Bull. geol. Soc. Am.*, **72**, 1283–4.

Milliman, J. D., and Emery, K. O. (1968). Sea levels during the past 35 000 years. *Science, N.Y.*, **162**, 1121–3.

Moe, D. (1970). The Post-glacial immigration of Picea abies into Fennoscandia. *Bot. Notiser*, **123**, 61–6.

Mörner, N.-A. (1969). Eustatic and climatic changes during the last 15 000 years. *Geologie Mijnb.*, **48**, 389–99.

Morrison, R. B. (1965). Quaternary geology of the Great Basin. *In* Wright and Frey, 1965, 265–85.

Morrison, R. B. (1968). Means of time-stratigraphic division and long-distance correlation of Quaternary successions. *In* Morrison and Wright, 1968, 1–113.

Morrison, R. B. (1969). The Pleistocene–Holocene boundary : an evaluation of the various criteria used for determining it on a provincial basis, and suggestions for establishing it world-wide. *Geologie Mijnb.*, **48**, 363–71.

Morrison, R. B., Gilluly, J., Richmond, G. M., and Hunt, C. B. (1957). In behalf of the Recent. *Am. J. Sci.*, **255**, 385–93.

Morrison, R. B., and Wright, H. E., Jr. (eds.) (1968). *Means of Correlation of Quaternary Successions*, University of Utah Press, Salt Lake City.

Mortimer, C., Clark, A. H., and Schufle, J. A. (1971). Ion exchange and radiocarbon dating of alluvial sediments from the lower Río Copiapó, Chile. *Nature, Lond.*, **229**, 54–5.

Munk, W. H., and MacDonald, G. J. F. (1960). *The Rotation of the Earth,* Cambridge University Press.

Nace, R. L. (1969). World Water inventory and control. *In Water, Earth, and Man* (ed. R. J. Chorley), Methuen, London, pp. 31–42.

Neustadt, M. I., and Gudelis, V. (1965). Holocene problems. *Rep. VI Int. Cong. Quat. (INQUA) Warsaw 1961, Lódz,* **1,** 467–77.

Newell, N. D., and Bloom, A. L. (1970). The Reef Flat and 'Two-meter eustatic terrace' of some Pacific atolls. *Bull. geol. Soc. Am.,* **81,** 1881–94.

Ninkovich, D., and Heezen, B. C. (1965). Santorini Tephra. *In Submarine Geology and Geophysics* (eds. W. F. Whittard and R. Bradshaw), Butterworth, London, pp. 413–53.

Oakley, K. P. (1964). *Frameworks for Dating Fossil Man,* Weidenfeld and Nicolson, London.

Oakley, K. P. (1969). Analytical methods of dating bones. *In* Brothwell and Higgs, 1969, 35–45.

Olausson, E. (1965). Evidence of climatic changes in North Atlantic deepsea cores, with remarks on isotopic paleotemperature analysis. *Progr. Oceanog.,* **3,** 221–52.

Olsson, I. U. (ed.) (1970). *Radiocarbon Variations and Absolute Chronology* (Nobel Syposium 12), Almqvist and Wiksell, Stockholm, and John Wiley and Sons, New York and London (1970).

Opdyke, N. D., Glass, B., Hays, J. D., and Foster, J. (1966). Palaeomagnetic study of Antarctic deep-sea cores. *Science, N.Y.,* **154,** 349–57.

Page, N. R. (1972). On the age of the Hoxnian interglacial. *Geol. J.,* **8,** 129–42.

Pearson, H. W. (1901). Oscillations in the sea-level. *Geol. Mag.,* **8,** 167–74.

Pennington, W. (1969). *The History of British Vegetation,* English University sities Press, London.

Phleger, F. B. (1960). *Ecology and Distribution of Recent Foraminifera,* John Hopkins Press, Baltimore.

Playfair, J. (1802). *Illustrations of the Huttonian Theory of the Earth,* Dover, New York.

Poulson, T. L., and White, W. B. (1969). The cave environment. *Science, N.Y.,* **165,** 971–81.

Raikes, R. (1967). *Water, Weather and Prehistory,* John Baker, London.

Rankama, K. (ed.) (1965). *The Quaternary I,* Interscience, London.

Rankama, K. (1970). Global Precambrian stratigraphy: background and principles. *Scientia, Bologna,* **105,** 382–421.

Reade, T. M. (1872). The Post-glacial geology and physiography of West Lancashire and the Mersey estuary. *Geol. Mag.,* **9,** 111–19.

Reichenbach, H. (1958). *Space and Time,* Dover, New York.

Renfrew, C. (1970). New configurations in Old World archaeology. *World Archaeology,* **2,** 199–211.

Richmond, G. M. (ed.) (1959). American Commission on Stratigraphic Nomenclature. Report 6 : Committee on Pleistocene, Application of Stratigraphic Classification and Nomenclature to the Quaternary. *Bull. Am. Ass. Petrol. Geol.,* **43,** 663–73.

Richmond, G. M. (1962). Morphostratigraphic units in Pleistocene stratigraphy. *Bull. Am. Ass. Petrol. Geol.,* **46,** 1520–1.

Ross, C. A. (1970). Concepts in Late Paleozoic correlations. *Spec. Pap. geol. Soc. Am.,* **124,** 7–36.

Rossignol, M. (1969). Sédimentation palynologique récente dans la Mer Morte. *Pollen et Spores*, 11, 17–38.

Ruhe, R. V. (1969). *Quaternary Landscapes in Iowa*, Iowa State University Press, Ames.

Ruhe, R. V., and Daniels, R. B. (1965). Landscape erosion—geologic and historic. *J. Soil Wat. Conserv.*, 20, 52–57.

Runcorn, S. K. (ed.) (1970). *Palaeogeophysics*, Academic Press, London.

Russel, B. (1953). *Mysticism and Logic*, Penguin Books, Harmondsworth.

Russell, R. J. (1958). Geological geomorphology. *Bull. geol. Soc. Am.*, 69, 1–22.

Sackett, J. R. (1968). Method and theory of Upper Paleolithic Archeology in southwestern France. *In* Binford and Binford, 1968, 61–83.

Sandford, K. S., and Arkell, W. J. (1933). Palaeolithic man and the Nile valley in Nubia and Upper Egypt. *Publ. Orient. Inst. Chicago*, 17.

Sawyer, J. S. (ed.) (1966). *World Climate from 8000 to 0 B.C.*, Royal Meteorological Society, London.

Scheidegger, A. E. (1963). *Principles of Geodynamics* (2nd edition), Springer-Verlag, Berlin.

Schindewolf, O. H. (1957). Comments on some stratigraphic terms. *Am. J. Sci.*, 255, 394–9.

Schmid, E. (1969). Cave sediments and prehistory. *In* Brothwell and Higgs, 1969, 151–166.

Schnable, J. E., and Goodell, H. G. (1968). Pleistocene–Recent stratigraphy, evolution, and development of the Apalachicola coast, Florida. *Spec. Pap. geol. Soc. Am.*, 112.

Schofield, J. C. (1967). Post Glacial sea-level maxima a function of salinity? *J. Geosci., Osaka City Univ.*, 10, 115–18.

Scholl, D. W., and Stuiver, M. (1967). Recent submergence of southern Florida: a comparison with adjacent coasts and other eustatic data. *Bull. geol. Soc. Am.*, 78, 437–54.

Schumm, S. A. (1965). Quaternary paleohydrology. *In* Wright and Frey, 1965, 783–94.

Schumm, S. A., and Lichty, R. W. (1965). Time, space and causality in geomorphology. *Am. J. Sci.*, 263, 110–19.

Sears, M. (ed.) (1967). The Quaternary history of the ocean basins. *Prog. Oceanog.*, 4.

Shackleton, N. J. (1967). Oxygen isotope analyses and Pleistocene temperatures re-assessed. *Nature, Lond.*, 215, 15–17.

Shackleton, N. J. (1969). The last interglacial in the marine and terrestrial records. *Proc. R. Soc.*, B, 174, 135–54.

Shackleton, N. J., and Turner, C. (1967). Correlation between marine and terrestrial Pleistocene successions. *Nature, Lond.*, 216, 1079–82.

Shaw, A. B. (1964). *Time in Stratigraphy*, McGraw-Hill, New York.

Shepard, F. P. (1970). Lagoonal topography of Caroline and Marshall Islands. *Bull. geol. Soc. Am.*, 81, 1905–14.

Shepard, F. P., and Curray, J. R. (1967). Carbon-14 determination of sea level changes in stable areas. *In* Sears, 1967, 283–91.

Shotton, F. W. (1972). An example of hard-water error in radiocarbon dating of vegetable matter. *Nature, Lond.*, 240, 460–1.

Simpson, G. G. (1963). Historical Science. *In* Albritton, 1963, 24–48.

Smalley, I. J. (1967). The subsidence of the North Sea Basin and the geomorphology of Britain. *Mercian Geologist*, 2, 267–78.

Smiley, T. L. (ed.) (1955). *Geochronology*, University of Arizona Press, Tucson.

Solecki, R. S., and Leroi-Gourhan, A. (1961). Palaeoclimatology and archaeology in the Near East. *Ann. N.Y. Acad. Sci.*, **95**, 729–39.

Suess, H. E. (1970). Bristlecone-pine calibration of the radiocarbon timescale 5200 B.C. to the present. *In* Olsson, 1970, 303–11.

Suggate, R. P. (1958–61). Time-stratigraphic subdivision of the Quaternary, as viewed from New Zealand. *Quaternaria*, **5**, 5–17.

Suggate, R. P., and West, R. G. (1967). The substitution of local stage names for Holocene and Post-glacial. *Quaternaria*, **9**, 245–6.

Sylvester-Bradley, P. C. (1967). Towards an international code of stratigraphic nomenclature. *In Essays in Paleontology and Stratigraphy* (eds. C. Teichert and E. L. Yochelson), University of Kansas Press, Lawrence, pp. 49–56.

Tchernov, E. (1968). *Succession of Rodent Faunas During the Upper Pleistocene of Israel*, Paul Parey, Berlin.

Teichert, C. (1958). Some biostratigraphical concepts. *Bull. geol. Soc. Am.*, **69**, 99–119.

Teilhard de Chardin, P. (1965). *The Appearance of Man*, Collins, London.

Thompson, D'A. W. (1961). *On Growth and Form* (abr. by J. T. Bonner), Cambridge University Press.

Thurber, D. L. (1972). Problems of dating non-woody material from continental environments. *In Calibration of Hominoid Evolution* (eds. W. W. Bishop and J. A. Miller), Scottish Academic Press, Edinburgh, pp. 1–17.

Toulmin, S., and Goodfield, J. (1967). *The Discovery of Time*, Penguin Books, Harmondsworth.

Tournal, M. (1833). General considerations on the phenomenon of bone coverns. *In Man's Discovery of his Past* (ed. R. F. Heizer, 1962), Prentice-Hall, Englewood Cliffs, pp. 72–82.

Turekian, K. K. (ed.) (1971). *The Late Cenozoic Glacial Ages*, Yale University Press, New Haven.

Vita-Finzi, C. (1964). Observations on the Late Quaternary of Jordan. *Palestine Expl. Q.*, **95**, 19–33.

Vita-Finzi, C. (1966). The Hasa Formation. *Man*, **1**, 386–90.

Vita-Finzi, C. (1967). Late Quaternary alluvial chronology of northern Algeria. *Man*, **2**, 205–15.

Vita-Finzi, C. (1968). The Rharbian Formation of Morocco. *Man*, **3**, 485–7.

Vita-Finzi, C. (1969a). Fluvial geology. *In* Brothwell and Higgs, 1969, 135–50.

Vita-Finzi, C. (1969b). *The Mediterranean Valleys*, Cambridge University Press.

Vita-Finzi, C. (1969c). Mediterranean monoglacialism? *Nature, Lond.*, **224**, 173.

Vita-Finzi, C. (1969d). Late Quaternary continental deposits of central and western Turkey. *Man*, **4**, 605–19.

Vita-Finzi, C. (1969e). Late Quaternary alluvial chronology of Iran. *Geol. Rdsch.*, **58**, 951–73.

Vita-Finzi, C. (1970a). Time, stratigraphy and the Quaternary. *Scientia, Bologna*, **105**, 725–36.

Vita-Finzi, C. (1970b). Alluvial history of central Mexico. *Nature, Lond.*, **227**, 596–7.

Vita-Finzi, C. (1971). Heredity and environment in clastic sediments: silt/clay depletion. *Bull. geol. Soc. Am.*, **82**, 187–90.

Vita-Finzi, C. (1973). Supply of fluvial sediment to the Mediterranean during the last 20 000 years. *In The Mediterranean* (ed. D. J. Stanley), in the press.

Vita-Finzi, C., and Dimbleby, G. W. (1971). Medieval pollen from Jordan. *Pollen et Spores*, **13**, 415–20.

Vita-Finzi, C., and Higgs, E. S. (1970). Prehistoric economy in the Mount Carmel area of Palestine: site catchment analysis. *Proc. prehist. Soc.*, **36**, 1–37.

Vita-Finzi, C., and Smalley, I. J. (1970). Weathering of quartz particles in alluvial red beds. *Naturwissenschaften*, **57**, 305.

Wager, L. R. (1964). The history of attempts to establish a quantitative time-scale. *In Harland et al.*, 1964, 13–28.

Walcott, R. I. (1972). Past sea levels, eustasy and deformation of the Earth. *Quaternary Research*, **2**, 1–14.

Wallén, C. C. (1956). Fluctuations and variability in Mexican rainfall. *In The Future of Arid Lands* (ed. G. F. White), *Amer. Ass. Adv. Sci., Washington*, pp. 141–155.

Wasserburg, G. J. (1966). Geochronology, and isotopic data bearing on development of the continental crust. *In Advances in Earth Science* (ed. P. M. Hurley), M.I.T. Press, Cambridge, Mass., pp. 431–59.

Waterbolk, H. T. (1971). Working with radiocarbon dates. *Proc. prehist. Soc. Lond.*, **37**, part II, 15–33.

Watson, R. A. (1966). Is geology different?: a critical discussion of 'The Fabric of Geology'. *Philosophy Sci.*, **33**, 172–85.

Wellman, H. W. (1955). New Zealand Quaternary tectonics. *Geol. Rdsch.*, **43**, 248–57.

West, R. G. (1968). *Pleistocene Geology and Biology*, Longmans, London.

Wheeler, M. (1954). *Archaeology from the Earth*, Clarendon Press, Oxford.

Whitrow, G. J. (1961). *The Natural Philosophy of Time*, Nelson, London.

Wickman, F. E. (1968). How to express time in geology. *Am. J. Sci.*, **266**, 316–18.

Wiener, C. (1965). *Cybernetics* (2nd edition), M.I.T. Press, Cambridge, Mass.

Wiggers, A. J. (1954). Compaction of sediments older than Holocene in relation to the subsidence of the Netherlands. *Geologie Mijnb.*, **16**, 179–84.

Williams, G. (1970). Piedmont sedimentation and late Quaternary chronology in the Biskra region of the northern Sahara. *Z. geomorph. Suppl.*, **10**, 40–63.

Woldstedt, P. (1967). The Quaternary of Germany. *In The Quaternary 2* (ed. K. Rankama), *Interscience, London*, pp. 239–300.

Woodward, H. B. (1891). Report on the Pliocene, Pleistocene and Recent. *Proc. 4th Int. Geol. Cong. 1888*, London, B 19–38.

Wright, H. E., Jr. (1971). Late Quaternary vegetational history of North America. *In Turekian*, 1971, 425–64.

Wright, H. E., Jr., and Frey, D. G. (eds.) (1965). *The Quaternary of the United States*, Princeton University Press, Princeton.

Wyllie, P. J. (1971). *The Dynamic Earth*, Wiley, New York.

Zak, I., and Freund, R. (1966). Recent strike slip movements along the Dead Sea Rift. *Israel J. Earth-Sci.*, **15**, 33–7.

Zeuner, F. E. (1958). *Dating the Past* (4th edition), Methuen, London.

Zeuner, F. E. (1959). *The Pleistocene Period*, Hutchinson, London.

Index